T5-CAG-993

Springer Tracts in Modern Physics
Volume 138

Springer
Berlin
Heidelberg
New York
Barcelona
Budapest
Hong Kong
London
Milan
Paris
Santa Clara
Singapore
Tokyo

Springer Tracts in Modern Physics

Covering reviews with emphasis on the fields of Elementary Particle Physics, Solid-State Physics, Complex Systems, and Fundamental Astrophysics

Manuscripts for publication should be addressed to the editor mainly responsible for the field concerned:

Gerhard Höhler
Institut für Theoretische Teilchenphysik
Universität Karlsruhe
Postfach 6980
D-76128 Karlsruhe
Germany
Fax: +49 (7 21) 37 07 26
Phone: +49 (7 21) 6 08 33 75
Email: hoehler@fphvax.physik.uni-karlsruhe.de

Johann Kühn
Institut für Theoretische Teilchenphysik
Universität Karlsruhe
Postfach 6980
D-76128 Karlsruhe
Germany
Fax: +49 (7 21) 37 07 26
Phone: +49 (7 21) 6 08 33 72
Email: johann.kuehn@physik.uni-karlsruhe.de

Thomas Müller
IEKP
Fakultät für Physik
Universität Karlsruhe
Postfach 6980
D-76128 Karlsruhe
Germany
Fax:+49 (7 21) 6 07 26 21
Phone: +49 (7 21) 6 08 35 24
Email: mullerth@vxcern.cern.ch

Roberto Peccei
Department of Physics
University of California, Los Angeles
405 Hilgard Avenue
Los Angeles, California 90024-1547
USA
Fax: +1 310 825 9368
Phone: +1 310 825 1042
Email: robertop@college.ucla.edu

Frank Steiner
Abteilung für Theoretische Physik
Universität Ulm
Albert-Einstein-Allee 11
D-89069 Ulm
Germany
Fax: +49 (7 31) 5 02 29 24
Phone: +49 (7 31) 5 02 29 10
Email: steiner@physik.uni-ulm.de

Joachim Trümper
Max-Planck-Institut
für Extraterrestrische Physik
Postfach 1603
D-85740 Garching
Germany
Fax: +49 (89) 32 99 35 69
Phone: +49 (89) 32 99 35 59
Email: jtrumper@mpe-garching.mpg.de

Peter Wölfle
Institut für Theorie
der Kondensierten Materie
Universität Karlsruhe
Postfach 69 80
D-76128 Karlsruhe
Germany
Fax: +49 (7 21) 69 81 50
Phone: +49 (7 21) 6 08 35 90/33 67
Email: woelfle@tkm.physik.uni-karlsruhe.de

Martin Erdmann

The Partonic Structure of the Photon

Photoproduction at the Lepton-Proton Collider HERA

With 59 Figures

Springer

Dr. Martin Erdmann

Universität Heidelberg
Physikalisches Institut
Philosophenweg 12
D-69120 Heidelberg
E-mail: erdmann@physi.uni-heidelberg.de

Cataloging-in-Publication Data applied for

Die Deutsche Bibliothek - CIP-Einheitsaufnahme

Erdmann, Martin:
The partonic structure of the photon : photoproduction at the lepton-proton collider HERA / Martin Erdmann. - Berlin ; Heidelberg ; New York ; Barcelona ; Budapest ; Hong Kong ; London ; Milan ; Paris ; Santa Clara ; Singapore ; Tokyo : Springer, 1997
(Springer tracts in modern physics ; Vol. 138)
ISBN 3-540-62621-2

Physics and Astronomy Classification Scheme (PACS):
14.70.Bh, 12.38.-t, 12.38.Qk, 13.60.-r, 13.85.Hd, 13.87.-a

ISBN 3-540-62621-2 Springer-Verlag Berlin Heidelberg New York

Typesetting: Camera-ready copy by the author using a Springer TEX macro-package
Cover design: *design & production* GmbH, Heidelberg
SPIN: 10546202 56/3144-5 4 3 2 1 0 – Printed on acid-free paper

For Georg Wittich

Preface

The photon is the exchange particle of electromagnetic interactions. However, at high energies $E_\gamma \geq O(1)$ GeV, the photon can fluctuate into a quark–anti-quark pair: like a hadron, the photon then takes part in strong interaction processes. These processes are described by quantum chromodynamics (QCD). Note that QCD predicts remarkable differences between a photon and a hadron and their interactions with a target particle, e.g., a proton.

In 1992, with the setting up of the electron–proton collider HERA at DESY in Hamburg, interactions of almost real photons could be studied in many aspects, some of them for the first time. The large photon–proton center-of-mass energies allow for quantitative tests of perturbative QCD and for quantitative studies of the partons coming from the photon.

This book summarizes the HERA results on the partonic interactions of the photon and compares them with QCD predictions and results from fixed-target photon–nucleon, two–photon and proton–anti-proton experiments. It has been written as an experimental review including the basic theoretical aspects for the understanding of the data, addressing theoreticians and experimentalists working in elementary particle physics. I hope it will be a useful consultative work for physicists working on the γp experiments at HERA and FNAL, on the $\gamma\gamma$ experiments at CERN and KEK, and for physicists studying the jet underlying event energy at the FNAL $\overline{\text{p}}$p experiments. Advanced students who have heard lectures on elementary particle physics may find here an introduction to contemporary research on the photon.

I wish to thank very much F. Eisele and Ch. Berger for numerous fruitful discussions on the interpretation of the photoproduction data and the analysis strategy. I thank all members of the H1 and ZEUS photoproduction groups for many important contributions. Special thanks go to M. Arpagaus, A. Buniatian, M. Colombo, A. DeRoeck, M. Gebauer, W. Hoprich, H. Hufnagel, R. Kaschowitz, G. Knies, H. Kolanoski, S. Levonian, P. O. Meyer, P. Pfeiffenschneider, H. Rick, A. Rostovtsev, C. Schwanenberger, U. Siewert, M. Steenbock, and W. Walkowiak. I have benefitted from interesting discussions with P. Aurenche, J. Binnewies, J. Chyla, M. Drees, P. Hoyer, M. Klasen, B. Kniehl, G. Kramer, S. G. Salesch, G. A. Schuler, T. Sjöstrand, and A. Vogt. For constructive comments on the contents of this review,

I thank S. Egli, F. Eisele, W. Gradl, S. Maxfield, H. Rick, P. Schleper, G. Schuler, S. Söldner-Rembold and M. Werner.

I wish to thank H.J. Koelsch and Th. Müller for their support in publishing this review in the STMP Springer series. I enjoyed working with the Springer team and thank U. Heuser, J. Lenz, F. Meyer, and V. Wicks for helping to create the appearance of this book. Special thanks go to S. Maxfield and C. Simpson for the English corrections, and to W. Gradl who adapted the original LaTex file to the Springer design.

Heidelberg, March 1997 *Martin Erdmann*

Contents

1. Introduction 1

2. Predictions for Hard Photon Interactions 5
2.1 Photons 5
2.1.1 Time of $e \to e\gamma$ and $\gamma \to q\overline{q}$ fluctuations 6
2.1.2 The Equivalent Photon Approximation 8
2.1.3 The Photon Structure Function F_2^γ 10
2.1.4 The Vector Meson Dominance Model 12
Summary 12
2.2 The QCD Description of Photon Interactions 12
2.2.1 QCD Predictions for Parton Scattering Processes 13
2.2.2 Leading Order QCD Predictions for Jet and Particle Cross Sections 16
2.2.3 Four Observables to Determine the Parton Kinematics 20
2.2.4 Next-to-Leading Order QCD Predictions 21
2.2.5 Multiple Parton Interactions 22
2.2.6 Event Generators 25
Summary 28

3. Two-γ Physics: Deep Inelastic Lepton–Photon Scattering 29
3.1 Measurements of the Photon Structure Function F_2^γ 29
Summary 31
3.2 Parameterizations of the Parton Distributions in the Photon 32
Summary 36

4. Photon–Proton Interactions at HERA 37
4.1 Introduction 37
4.1.1 Photoproduction Processes 37
4.1.2 Event Kinematics 38
4.1.3 The HERA Accelerator 41
4.1.4 The H1 and ZEUS Experiments 42
Summary 48

4.2 Verification of QCD Predictions in γp Scattering 48
4.2.1 Hard Scattering Processes: Particle Production 49
4.2.2 Hard Scattering Processes: Jet Production 51
4.2.3 Resolved Photon Interactions 59
4.2.4 Direct Photon Interactions 63
4.2.5 The Parton Scattering Angle 66
4.2.6 Higher-Order QCD Effects 68
Summary 68
4.3 Multiple Parton Interactions 70
4.3.1 Double Parton Scattering in pp and $\overline{\text{p}}$p Collisions 71
4.3.2 Inclusive Transverse Energy Cross Section 72
4.3.3 Transverse Energy Flow 75
4.3.4 Energy–Energy Correlations 76
4.3.5 Underlying Event Energy 77
Summary 81
4.4 Parton Distributions of the Photon 82
4.4.1 Lower Limit on the Parton Fractional Energy 82
4.4.2 Procedures to Extract the Parton Distributions of the Photon 83
4.4.3 Particle Cross Sections 84
4.4.4 Single Differential Jet Cross Sections 86
4.4.5 Measurement of the Gluon Distribution 92
4.4.6 Double Differential Di-Jet Cross Section 95
4.4.7 Measurement of the Effective Parton Distribution 97
Summary 99

5. Status and Future 101
5.1 Status: Hard Photoproduction at HERA 101
5.2 Prospects of High Energy Photon Interactions 103

References 105

Abbreviations and Variables 109

Index 113

1. Introduction

The interaction of high energy photons with leptons is very well described by the electro-weak theory. The interaction with hadrons, on the other hand, is full of surprises and has developed into an interesting field of QCD.

The first generation of photon–nucleon fixed-target scattering experiments revealed that, to a very good approximation, the photon behaves like a hadron that is quantitatively summarized by the vector meson dominance model (VDM) (for a review refer to [10]). In this model, the photon has a well-defined probability of fluctuating into vector mesons with the quantum numbers of the photon, spin= 1, negative parity and negative charge conjugation, e.g., $\varrho, \omega, \phi, \ldots$ (Fig. 1.1: VDM). The VDM gives relations between photon–nucleon and meson–nucleon reactions, which are well satisfied in interactions with small transverse energy in the final state.

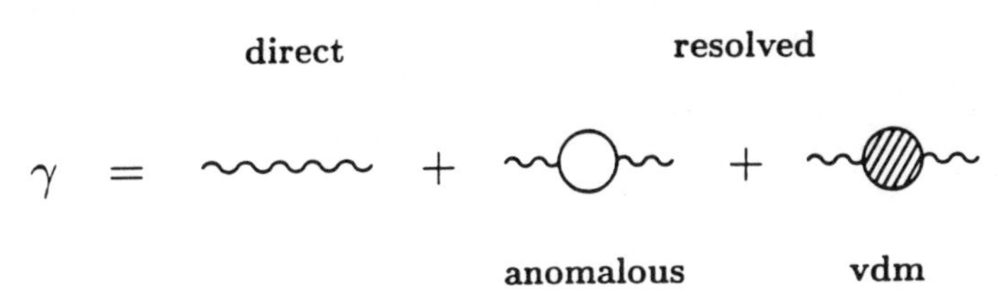

Fig. 1.1. Different states of the photon are shown: apart from the bare photon state (*direct*), the photon can fluctuate into quark–anti-quark pairs without forming a hadronic bound state (*anomalous*), or form a vector meson (*VDM*). The photon can therefore interact directly, or through its *resolved* states

The advent of QCD theory gave interesting and significant modifications to the VDM and its predictions. QCD predicts processes where the photon couples directly with quarks, leading to hard parton scattering and jets in the final state (Figs. 1.1, 1.2). The measurement of these processes was the goal of several fixed-target experiments at CERN and FNAL with center-of-mass (CM) energies up to $\sqrt{s_{\gamma \mathrm{p}}} = 27$ GeV (for a review refer to [94]). They found significant deviations from the VDM model by observing an excess of final-state hadrons with large transverse momenta, which can be quantitatively explained by direct photon–nucleon interactions. Also in recent years, evidence for jet structure in the final state has been reported [42]. The center-of-mass energy was, however, too small to make definite QCD tests.

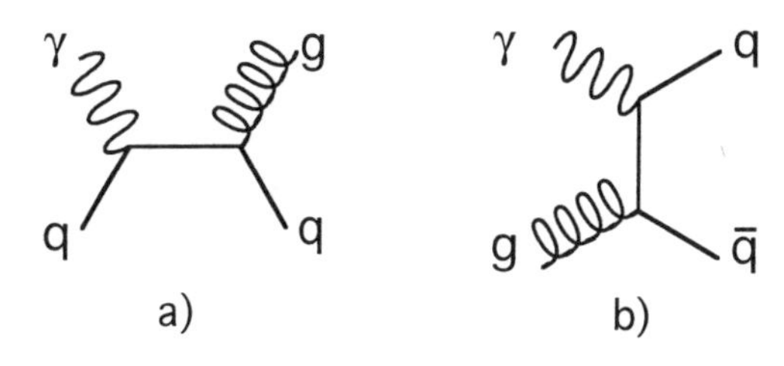

Fig. 1.2. Diagrams of direct photon–nucleon interactions are shown in leading order QCD: (**a**) the QCD Compton and (**b**) the photon-gluon-fusion processes

A second QCD prediction is that at sufficiently high transverse energy $q\bar{q}$ fluctuations of the photon without the formation of a hadronic bound state will dominate the vector meson contribution (Fig. 1.1: anomalous). This contribution can be quantitatively calculated within perturbative QCD, e.g., the photon structure function for the anomalous part is predicted. The existence of the anomalous coupling was first shown in measurements of the photon structure function. This was achieved in two-photon reactions at the e^+e^- colliders (Fig. 1.3) at PETRA/PEP (for reviews refer to [14, 80, 81]) and recently at TRISTAN and LEP. Hard scattering reactions involving resolved photons were observed in the interactions of two quasi-real photons by the measurement of particles with large transverse momentum and jet formation (TRISTAN and LEP).

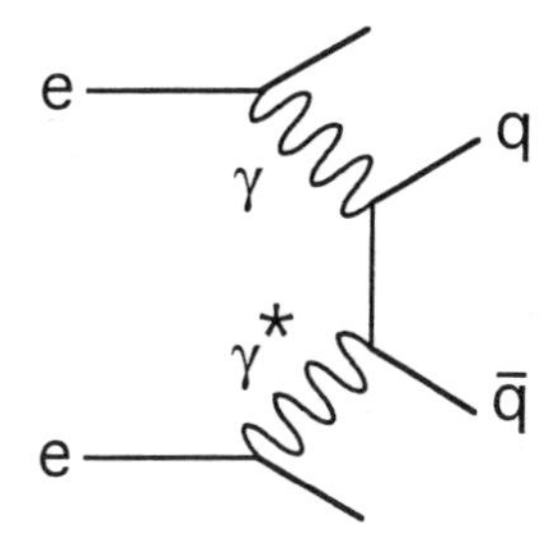

Fig. 1.3. Deep inelastic electron–photon scattering: a highly virtual photon γ^* ($Q^2 \gg 1$ GeV^2) probes the quark content of a quasi-real photon γ ($Q^2 \approx 0$ GeV^2)

The new HERA ep collider has been operational since 1992. This facility offers the possibility of studying photon–proton interactions at center-of-mass energies up to 300 GeV, which is an order of magnitude larger than in fixed-target experiments so far (Fig. 1.4). At these energies, the total γp cross section is still dominated by processes where the photon fluctuates into a vector meson (Fig. 1.5). A fraction of the events shows hard γp scattering processes, which manifest themselves in jets with high transverse energies. Both the direct photon and the resolved photon interactions can be extensively studied

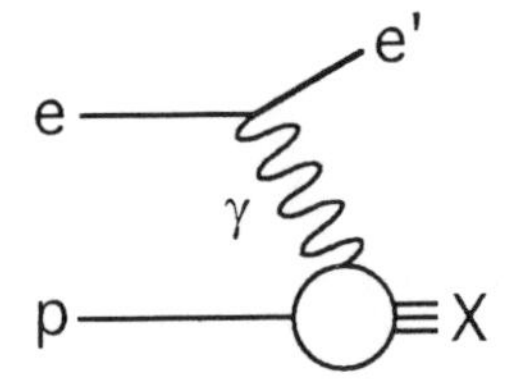

Fig. 1.4. The photoproduction regime in lepton–proton scattering corresponds to quasi-real photons ($Q^2 \approx 0\ \mathrm{GeV}^2$) interacting with protons

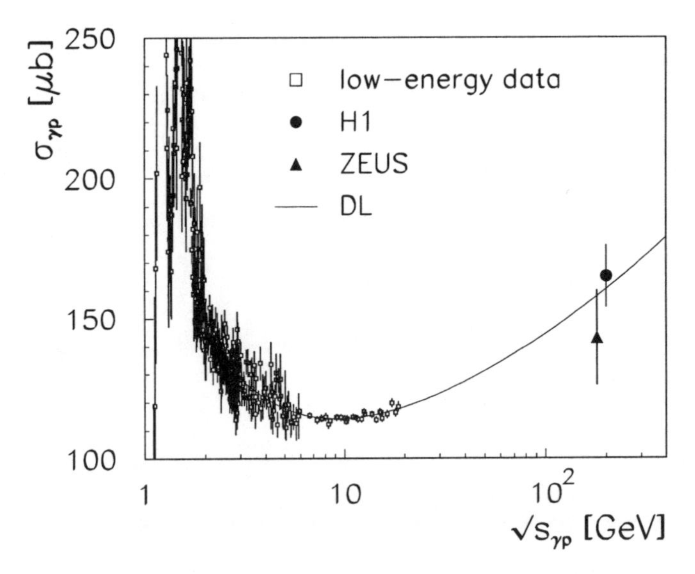

Fig. 1.5. Measurements of the total photon–proton cross section at different center-of-mass (CM) energies (*full circle*: H1 [67], *full triangle*: ZEUS Collab. [132], *open squares*: fixed-target data [13]). The energy dependence is compatible with the same moderate rise towards high CM energies as found in hadron-hadron total cross section measurements (*full curve* from [38])

and quantitatively compared with QCD predictions. The parton content of the photon – gluons and quarks – can be probed.

This review summarizes the HERA results from the first 5 years on the parton interactions of quasi-real photons and compares them with the results from other experiments.

2. Predictions for Hard Photon Interactions

In this section we want to introduce predictions for photon–target interactions that involve constituents of the photon and a target particle, which can be, for example, a nucleon or another photon. In current experiments, the photons are produced by highly energetic leptons. The photons carry fractional energies

$$0 < y \equiv \frac{\gamma\tau}{e\tau} = 1 - \frac{E_{e'}}{E_e} \cos^2\left(\frac{\theta_{e'}}{2}\right) < 1 . \tag{2.1}$$

Here e and τ are the four vectors of the beam lepton and the target particle, and γ is the photon four vector (Figs. 1.3, 1.4). The lepton beam energy is denoted by E_e, and $E_{e'}$ ($\theta_{e'}$) is the energy (angle) of the scattered lepton. The squared four momentum of the photon is negative and can be calculated from

$$0 < -\gamma^2 = -(e - e')^2 \equiv Q^2 = 4\, E_e E_{e'} \sin^2\left(\frac{\theta_{e'}}{2}\right) . \tag{2.2}$$

Q^2 is the virtuality of the photon. The topics we want to discuss in the following sections are:

1. How many photons are radiated by the electrons?
2. The Heisenberg uncertainty relation allows for fluctuations of photons into quark–anti-quark pairs. How long do these resolved photon states last?
3. How many partons are predicted to come from the photon?
4. What are the predictions for the direct and resolved photon contributions?
5. Which predictions exist for particle and jet production in photon–target interactions?

2.1 Photons

We first concentrate on questions 1–3 of the previous paragraph, which concern the radiation of photons from leptons and predictions on the partons coming from the photon.

2.1.1 Time of $\mathrm{e} \to \mathrm{e}\gamma$ and $\gamma \to \mathrm{q}\overline{\mathrm{q}}$ fluctuations

The typical time of a quark–anti-quark fluctuation of a highly energetic real photon in the rest frame of a target nucleon is large compared with the size of the nucleon. At $\mathrm{e}^+\mathrm{e}^-$ and ep colliders the photons are emitted by the leptons, which implies a two-step process for the creation of a $\mathrm{q}\overline{\mathrm{q}}$ pair: first, the electron fluctuates into an electron–photon state $\mathrm{e}\gamma$, then the photon fluctuates into a $\mathrm{q}\overline{\mathrm{q}}$ pair. As a result of the Heisenberg uncertainty relation,[1]

$$\Delta t \approx 1/\Delta E \, , \tag{2.3}$$

the lifetime $t(\mathrm{e} \to \mathrm{e}\gamma)$ of the electron–photon state is required to be larger than that of the $\gamma \to \mathrm{q}\overline{\mathrm{q}}$ fluctuation.

In the target rest frame, the time of fluctuation can be calculated from the uncertainty relation and relativistic kinematics (for a review see, for example, [72]). The energy difference between the initial state i with energy E_i and the state f of the fluctuation with constituents j and total energy E_f is given by

$$E_f - E_i = \frac{1}{2E_i} \left(\sum_j \frac{m_j^2 + p_{t,j}^2}{z_j} - m_i^2 \right) , \tag{2.4}$$

where longitudinal particle momenta are large compared to the transverse momenta. m_i and m_j are the masses of the initial particle and of the constituents. The transverse momenta $p_{t,j}$ of the constituents and their fractional energies z_j are measured with respect to the initial-state particle. In the following we discuss several applications of (2.4):

1. By emission of a highly virtual photon ($Q^2 \gg 1\ \mathrm{GeV}^2$) from an electron, transverse momenta are generated according to $p_{t,\gamma}^2 = Q^2(1-y)$. Such $\mathrm{e}\gamma$ fluctuations require the energy

$$E_{\mathrm{e}\gamma} - E_{\mathrm{e}} = \frac{1}{2E_\gamma} Q^2 \, . \tag{2.5}$$

The corresponding time of fluctuation for $E_\gamma = 1$ TeV and $Q^2 = 400\ \mathrm{GeV}^2$ is $t(\mathrm{e} \to \mathrm{e}\gamma) = 1$ fm (Fig. 2.1a: full curve).[2]

2. The emission of a quasi-real photon (virtuality $Q^2 \approx 0\ \mathrm{GeV}^2$) by an electron involves no transverse momentum in the collinear limit. The energy requirement for the $\mathrm{e}\gamma$ fluctuation is

$$E_{\mathrm{e}\gamma} - E_{\mathrm{e}} = \frac{1}{2E_\gamma} \frac{m_{\mathrm{e}}^2 y^2}{1-y} \, , \tag{2.6}$$

[1] The convention $\hbar = c = 1$ is used for convenience.

[2] In deep inelastic electron–nucleon scattering, the time of an $\mathrm{e}\gamma$ state can also be expressed by the momentum fraction x_{Bj} of the struck parton of the proton. Using $x_{\mathrm{Bj}} = Q^2/(2m_{\mathrm{N}}E_\gamma)$ results in $t(\mathrm{e} \to \mathrm{e}\gamma) = (m_{\mathrm{N}} x_{\mathrm{Bj}})^{-1} \equiv 2L_{\mathrm{i}}$, where m_{N} is the mass of the nucleon, and L_{i} is the so-called Ioffe length. A parton momentum fraction $x_{\mathrm{Bj}} = 0.2$ corresponds to $t(\mathrm{e} \to \mathrm{e}\gamma) = 1$ fm. Within this time, a subsequent fluctuation $\gamma \to \mathrm{q}\overline{\mathrm{q}}$ can happen.

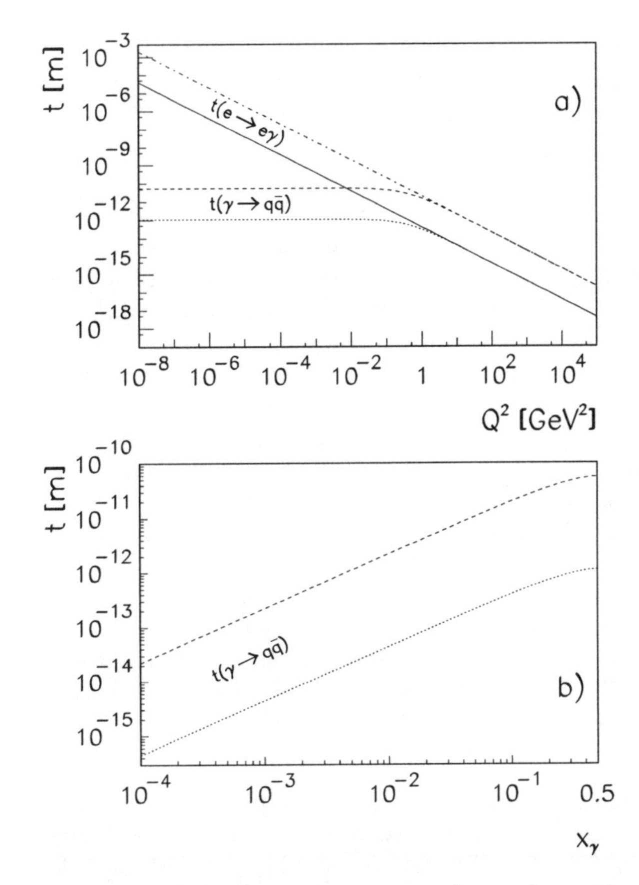

Fig. 2.1. (**a**) The time of an $\mathrm{e} \to \mathrm{e}\gamma$ fluctuation, as measured in the target rest frame, is shown as a function of the virtuality Q^2 of the photon for the photon energies $E_\gamma = 1$ TeV (*full curve*) and $E_\gamma = 50$ TeV (*dash-dotted curve*). The *dotted* (*dashed*) *curve* indicates the time of a subsequent $\gamma \to \mathrm{q}\overline{\mathrm{q}}$ fluctuation for $E_\gamma = 1$ TeV (50 TeV) with a symmetric energy sharing between the $\mathrm{q}\overline{\mathrm{q}}$ pair and $m_\mathrm{q}^2 + p_{t,\mathrm{q}}^2 = \Lambda_{\mathrm{QCD}}^2$. (**b**) The time of a $\gamma \to \mathrm{q}\overline{\mathrm{q}}$ fluctuation depends on the energy fraction x_γ carried by the quark: the *dotted* (*dashed*) *curve* represents the fluctuation of a quasi-real photon, i.e., $Q^2 \approx 0$, into a $\mathrm{q}\overline{\mathrm{q}}$ pair at the photon energy $E_\gamma = 1$ TeV (50 TeV). Here again we set $m_\mathrm{q}^2 + p_{t,\mathrm{q}}^2 = \Lambda_{\mathrm{QCD}}^2$

where E_e is the initial energy of the electron with mass m_e, and y is the energy fraction carried by the photon, which is here $y = E_\gamma / E_\mathrm{e}$. The ratio

$$Q_{\mathrm{min}}^2 \equiv \frac{m_\mathrm{e}^2 y^2}{1-y} \tag{2.7}$$

gives the smallest virtuality of photons that are generated by electrons: $Q_{\mathrm{min}}^2 \approx 10^{-7}\ \mathrm{GeV}^2$. At photon energies of $E_\gamma = 1$ TeV, the time of an $\mathrm{e}\gamma$ fluctuation is $t(\mathrm{e} \to \mathrm{e}\gamma) \approx 4\ \mu\mathrm{m}$ (Fig. 2.1a: full curve).

3. Fluctuations of the photon into a $\mathrm{q\overline{q}}$ pair depend on the energy fraction x_γ, which is carried by the quark relative to the photon energy E_γ. For quasi-real photons the required energy is

$$E_{\mathrm{q\overline{q}}} - E_\gamma = \frac{1}{2E_\gamma} \frac{m_{\mathrm{q}}^2 + p_{t,\mathrm{q}}^2}{x_\gamma(1-x_\gamma)} , \tag{2.8}$$

where m_{q} and $p_{t,\mathrm{q}}$ are the mass and transverse momentum of the (anti-) quark (Fig. 2.1b). Symmetric configurations between the quark and anti-quark, i.e., $x_\gamma = 0.5$, give the longest lifetime for such a $\mathrm{q\overline{q}}$ state. Using light quarks with small $p_{t,\mathrm{q}}$, e.g., setting $m_{\mathrm{q}}^2 + p_{t,\mathrm{q}}^2 = \Lambda_{\mathrm{QCD}}^2$, and a symmetric energy sharing at $E_\gamma = 1$ TeV results in a time $t(\gamma \to \mathrm{q\overline{q}}) = 1000$ fm for the fluctuation [dotted curves in Fig. 2.1: a) $Q^2 < 10^{-2}$, b) $x_\gamma = 0.5$]. As mentioned above, the formation of a $\mathrm{q\overline{q}}$ pair from an electron via a photon is only allowed if the time of the $\mathrm{q\overline{q}}$ fluctuation lies within the lifetime of the $\mathrm{e}\gamma$ state. At large $Q^2 \gg 1$ GeV^2, the time $t(\gamma \to \mathrm{q\overline{q}})$ is therefore limited by the time $t(\mathrm{e} \to \mathrm{e}\gamma)$ (Fig. 2.1a: dotted and full curves).
4. A subsequent formation of a gluon from the (anti-) quark $\mathrm{q} \to \mathrm{qg}$ has a lifetime which is suppressed by the energy fraction x_γ of the quark relative to the photon, and the quark energy fraction z, which is taken by the gluon: $x_\gamma \cdot z(1-z)$. Using $x_\gamma = 0.5$ and an asymmetric energy share between the quark and the massless gluon of $z = 0.1$ results in a lifetime which is 20 times shorter than the $\gamma \to \mathrm{q\overline{q}}$ fluctuation.

At HERA collisions between protons and quasi-real photons are studied at γ energies around 20 TeV, measured relative to the proton rest frame. Therefore $\mathrm{q\overline{q}}$ fluctuations from the photons typically last $t \sim 10^4$ fm. The time of fluctuation involving gluons is 1–2 orders of magnitude shorter than the quark fluctuations. Note that the time of photon fluctuations is finite. Therefore, we expect both direct and resolved photon interactions.

2.1.2 The Equivalent Photon Approximation

How many quasi-real photons are emitted by electrons, and what is the energy spectrum of these photons? These questions can be answered, e.g., by calculating electron–nucleon scattering in the equivalent photon approximation (EPA) (e.g., [23]). The field of a fast charged particle is similar to electromagnetic radiation. This radiation may be interpreted as a flux of photons with energy distribution $n(y)$ where y denotes the energy fraction of the photons relative to the electron energy. Electromagnetic electron–nucleon scattering can therefore be reduced to photon–nucleon interactions:

$$\mathrm{d}\sigma_{\mathrm{eN}}(y, Q^2) = \sigma_{\gamma\mathrm{N}}(y)\ \mathrm{d}n(y, Q^2) , \tag{2.9}$$

where $\sigma_{\gamma\mathrm{N}}$ is the total photo-absorption cross section, and Q^2 is the virtuality of the photons.

The first photon spectra were calculated by Weizsäcker and Williams [126, 127], neglecting the virtuality of the photon and terms involving the longitudinal photon polarization. This approximation is usually referred to as the Weizsäcker–Williams approximation (WWA). When the emission of quasi-real photons is integrated in a logarithmically large interval $Q^2_{\mathrm{min}} \leq Q^2 \leq Q^2_{\mathrm{max}} \ll 1$ GeV^2 and in a small energy bin dy, the equivalent number of photons is

$$\mathrm{d}n(y, Q^2_{\mathrm{max}}) = f_{\gamma/\mathrm{e}}(y, Q^2_{\mathrm{max}})\ \mathrm{d}y \tag{2.10}$$

with

$$f_{\gamma/\mathrm{e}} = \frac{\alpha}{2\pi}\left[\frac{1+(1-y)^2}{y}\ \ln\frac{Q^2_{\mathrm{max}}}{Q^2_{\mathrm{min}}} - 2m_{\mathrm{e}}^2 y\left(\frac{1}{Q^2_{\mathrm{min}}} - \frac{1}{Q^2_{\mathrm{max}}}\right)\right]. \tag{2.11}$$

Here α is the fine structure constant and Q^2_{min} is the kinematic lower limit shown in (2.7).

In Fig. 2.2 the energy spectrum of quasi-real photons, emitted by electrons, is shown for $Q^2_{\mathrm{max}} = 0.01$ GeV^2. The number of photons decreases at large photon energies $y > 0.2$ by an order of magnitude, but rises steeply towards small photon energies $y < 0.2$.

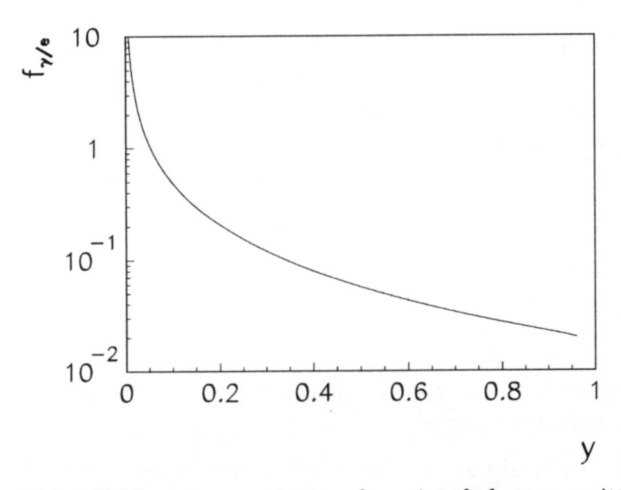

Fig. 2.2. The energy spectrum of quasi-real photons, emitted by electrons, is shown as a function of the scaled photon energy $y = E_\gamma/E_\mathrm{e}$ for a maximum virtuality $Q^2_{\mathrm{max}} = 0.01$ GeV^2

The accuracy of the WWA has been studied for many different processes (see e.g. [23]). In the case of photoproduction at HERA, the accuracy has been calculated, e.g., for the QED Compton process at large transverse photon momenta $p_t > 10$ GeV [11]. For events where the electron is detected at small scattering angles in the laboratory $1 \leq \theta \leq 10$ mrad, the WWA is better than 1%. For jet production with transverse jet energies $E_t^{\rm jet} \gg \sqrt{Q^2}$ and untagged electrons with $Q^2 < 4$ GeV^2, corrections to the WWA are at a level of a few percent [77].

2.1.3 The Photon Structure Function F_2^γ

The splitting of a photon into a quark–anti-quark pair can be calculated in the quark parton model (QPM). When a photon splits into a $\rm q\bar{q}$ pair, the quark carries an energy fraction x_γ, measured relative to the photon energy. Since the quark and anti-quark densities in the photon should be symmetric, and their fractional momenta are coupled in every process, a simple inverse relation holds for the probability of finding a quark in the photon $f_{\rm q/\gamma}$, and the probability of finding a photon in a quark $f_{\rm \gamma/q}$ [21, 33, 59]:

$$f_{\rm q/\gamma}(x_\gamma) = x_\gamma\ f_{\rm \gamma/q}(1/x_\gamma)\ . \tag{2.12}$$

The functional form of $f_{\rm \gamma/q}$ is the same as that of $f_{\rm \gamma/e}$ given in (2.11), scaled by the square of the quark charge $e_{\rm q}$ [ignoring the correction term $2m_{\rm e}^2 y(1/Q^2_{\rm min} - 1/Q^2_{\rm max})$]:

$$f_{\rm q/\gamma}(x_\gamma) = e_{\rm q}^2\ \frac{\alpha}{\pi}\ \left(x_\gamma^2 + (1-x_\gamma)^2\right)\ \ln\frac{Q^2}{m_{\rm q}^2}\ . \tag{2.13}$$

Here $m_{\rm q}$ is a measure of the mass of 'free' quarks. A measurement of the analogous QED process $f_{\mu/\gamma}$ resulted in a precise determination of the μ mass [116]. To compare with experiments, the probabilities $f_{\rm q/\gamma}$ are summed over all colors and flavors resulting in a prediction for the photon structure function F_2^γ:

$$F_2^\gamma(x_\gamma, Q^2) = x_\gamma \sum_{n_c, n_f} e_{\rm q}^2\ f_{\rm q/\gamma}(x_\gamma, Q^2)\ , \tag{2.14}$$

$$F_2^\gamma(x_\gamma, Q^2)\,[\mathrm{QPM}] = 3\sum_{n_f} e_{\rm q}^4\,\frac{\alpha}{\pi}\,x_\gamma\left(x_\gamma^2 + (1-x_\gamma)^2\right)\ \ln\frac{Q^2}{m_{\rm q}^2}\ . \tag{2.15}$$

This QPM photon structure function F_2^γ has three features that are different from hadronic structure functions:

1. The quark charges $e_{\rm q}$ contribute to the fourth power, compared with quadratic contributions in hadronic structure functions.
2. The photon structure function increases with increasing energy fraction x_γ of the quark from the photon.
3. The structure function of the quasi-real photon depends directly on the scale Q^2 at which it is probed by a highly virtual photon. In hadronic structure functions, Q^2 only enters via the QCD evolution equations.

QCD corrections to the simple QPM photon structure function can be calculated, e.g., from the DGLAP evolution equations. These evolution equations are inhomogeneous for the quark distributions of the photon to account for the pointlike coupling of the photon to quarks [first term on the right-hand side in (2.16)]:

$$\frac{\mathrm{d}f_{\mathrm{q}/\gamma}(x_\gamma,Q^2)}{\mathrm{d}\ln Q^2} = \frac{\alpha}{2\pi}\, e_{\mathrm{q}}^2 P_{\mathrm{q}\gamma}(x_\gamma) + \frac{\alpha_s}{2\pi} \times \left[P_{\mathrm{qq}}(x_\gamma) \otimes f_{\mathrm{q}/\gamma}(x_\gamma,Q^2) + P_{\mathrm{qg}}(x_\gamma) \otimes f_{\mathrm{g}/\gamma}(x_\gamma,Q^2)\right] \tag{2.16}$$

$$\frac{\mathrm{d}f_{\mathrm{g}/\gamma}(x_\gamma,Q^2)}{\mathrm{d}\ln Q^2} = \frac{\alpha_s}{2\pi}\Big[2P_{\mathrm{gq}}(x_\gamma) \otimes \sum_{n_f} f_{\mathrm{q}/\gamma}(x_\gamma,Q^2) + P_{\mathrm{gg}}(x_\gamma, n_f) \otimes f_{\mathrm{g}/\gamma}(x_\gamma,Q^2)\Big] . \tag{2.17}$$

Here P_{ij} denote the splitting functions, e.g., $P_{\mathrm{q}\gamma}$ gives the probability of the photon radiating a quark. The sum runs over all quark flavors n_f. The convolution integral is defined as $a(x_\gamma) \otimes b(x_\gamma) = \int_{x_\gamma}^{1} (\mathrm{d}y/y) a(x_\gamma/y) b(y)$.

It is non-trivial that these QCD corrections preserve the $\ln Q^2$ dependence of the QPM photon structure function (2.15) [128]. The leading order QCD prediction for the quark density in the photon is given by

$$f_{\mathrm{q}/\gamma}(x_\gamma) = e_{\mathrm{q}}^2\, \frac{\alpha}{\pi}\, \left(x_\gamma^2 + (1-x_\gamma)^2\right)\, \ln\frac{Q^2}{\Lambda_{\mathrm{QCD}}^2} . \tag{2.18}$$

The corresponding expression for the photon structure function is

$$F_2^\gamma(x_\gamma, Q^2) = 3\sum_{n_f} e_{\mathrm{q}}^4\, \frac{\alpha}{\pi}\, x_\gamma\, \left(x_\gamma^2 + (1-x_\gamma)^2\right)\, \ln\frac{Q^2}{\Lambda_{\mathrm{QCD}}^2} . \tag{2.19}$$

Equation (2.19) accounts for the pointlike/anomalous photon contribution. Note that possible bound states between the quark and the anti-quark are not included.

Since the strong coupling constant is to first order $\alpha_s \propto \left(\ln\left(Q^2/\Lambda_{\mathrm{QCD}}^2\right)\right)^{-1}$, the photon structure function (2.19) is proportional to the ratio of the electromagnetic and strong coupling constants:

$$F_2^\gamma(x_\gamma, Q^2) \propto \frac{\alpha}{\alpha_s} . \tag{2.20}$$

The photon structure function F_2^γ can be directly measured by deep inelastic electron–photon scattering experiments, which could, in principle, give a precise determination of the QCD parameter Λ_{QCD}, or α_s, respectively. Their measurements show that the anomalous photon processes cannot be separated from nonperturbative effects. The results of the experiments will be discussed in Sect. 3.

2.1.4 The Vector Meson Dominance Model

A very different ansatz to describe the photon and its interactions was postulated in the so-called vector meson dominance model (VDM): photons are essentially in a hadronic state when they interact with a target particle $\mathcal{T}$. The photoproduction cross sections were set proportional to the sum of vector-meson–target cross sections [99]:

$$\sigma_{\gamma\mathcal{T}} = \sum_{\mathrm{V}=\varrho,\omega,\phi\ldots} \frac{\pi\,\alpha}{\gamma_{\mathrm{V}}^2}\,\sigma_{\mathrm{V}\mathcal{T}}\;. \tag{2.21}$$

The photon cross sections are suppressed by α. The photon–vector-meson coupling constants γ_{V} can be extracted from meson decay experiments.

The VDM model successfully describes photon–target interactions which are dominated by soft collisions, e.g., total cross section measurements. In hard interactions involving photons, the label 'VDM' is synonymous with that part of resolved photon interactions where the quark–anti-quark pair of the photon forms a hadronic bound state before the scattering process.

Summary

1. At the ep and $\mathrm{e}^+\mathrm{e}^-$ colliders, quasi-real photons are emitted by the leptons where the rate of these photons can be calculated with good accuracy by the equivalent photon approximation.
2. In the target rest frame, the time of a $\gamma \to \mathrm{q}\overline{\mathrm{q}}$ fluctuation is very large compared to the size of a nucleon. Such partonic states lead to resolved photon interactions.
3. Two extreme views of the resolved photon exist:
   a) The photon fluctuates into a hadron (VDM).
   b) The quark and anti-quark propagate without forming a hadronic bound state: perturbative QCD predicts the distributions of quarks and gluons in the photon (anomalous resolved photon).
4. Since the time of $\gamma \to \mathrm{q}\overline{\mathrm{q}}$ fluctuations is finite, direct interactions of the photons with partons of a target particle should exist.

2.2 The QCD Description of Photon Interactions

QCD theory predicts the partonic interactions between a photon and a target particle. In the following, we want to list the leading order matrix elements and summarize the predictions for jet and particle cross sections in leading and next-to-leading order.

2.2.1 QCD Predictions for Parton Scattering Processes

The cross sections of elastic parton–parton scattering processes are predicted by QCD theory. For given initial-state parton energies E_1, E_2, the distribution of the scattering angle $\hat{\theta}$ in the parton center-of-mass system is calculated (Fig. 2.3). The results are usually presented in terms of the Lorentz invariant Mandelstam variables $\hat{s}, \hat{t}, \hat{u}$ which are, for massless partons, directly related to the parton energies and the scattering angle:

$$\hat{s} = 4E_1E_2 , \tag{2.22}$$

$$\hat{t} = -\frac{\hat{s}}{2}\left(1 - \cos\hat{\theta}\right) , \tag{2.23}$$

$$\hat{u} = -\frac{\hat{s}}{2}\left(1 + \cos\hat{\theta}\right) . \tag{2.24}$$

The differential parton cross sections can be written in the form

$$\frac{d\hat{\sigma}}{d\hat{t}} = \frac{|M|^2}{16\pi\ \hat{s}^2} , \tag{2.25}$$

with the matrix elements M of massless partons listed in Table 2.1 [80, 84]. The corresponding Feynman diagrams are shown in Figs. 1.2, 1.3, 2.4. Only two of the Mandelstam variables are independent of each other ($\hat{s}+\hat{t}+\hat{u} = 0$). For example, for a given parton center-of-mass energy $\sqrt{\hat{s}}$, the cross sections vary only with the parton scattering angle $\hat{\theta}$.

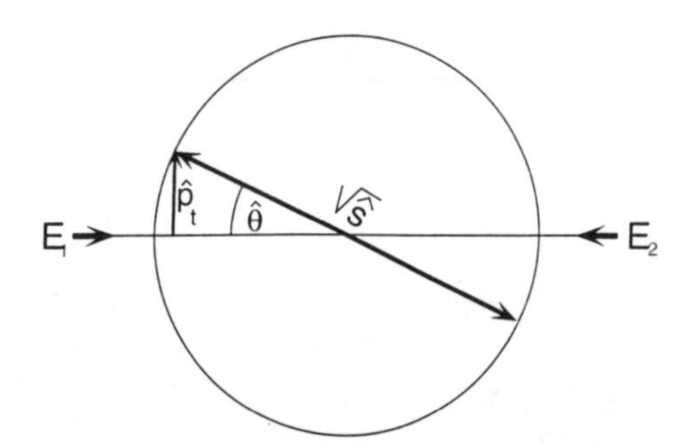

Fig. 2.3. An elastic parton scattering process is shown in the parton–parton center-of-mass system. The center-of-mass energy is denoted by $\sqrt{\hat{s}}$, the scattering angle is $\hat{\theta}$, and $\hat{p}_t$ is the transverse momentum of the scattered partons

In Fig. 2.5 the matrix elements are shown as a function of $\cos\hat{\theta}$. In the $\hat{\theta} = 90°$ scattering region, the matrix elements differ by three orders of magnitude. The dominant parton cross section results from gluon–gluon scattering gg $\to$ gg. Most of the matrix elements diverge at $\cos\hat{\theta} = 1$ (-1), which corresponds to small-angle forward (backward) scattering.

Since forward and backward scattering cannot be distinguished by the experiments in practice, only the absolute value of $\cos\hat{\theta}$ is relevant. In Fig. 2.6 the shapes of the matrix elements are shown as a function of $|\cos\hat{\theta}|$ for the direct photon processes and the resolved photon processes with the largest

Table 2.1. Leading order QCD matrix elements are listed for resolved (*upper 8*) and direct (*lower 3*) photon–nucleon and two-photon-scattering processes. The matrix elements for direct interactions are the QCD Compton, the photon-gluon fusion, and the two-photon process

process	$\|M\|^2/\pi^2$
$\mathrm{qq}' \to \mathrm{qq}' = \mathrm{q\bar{q}}' \to \mathrm{q\bar{q}}'$	$\frac{64}{9}\alpha_s^2\left(\frac{\hat{s}^2+\hat{u}^2}{\hat{t}^2}\right)$
$\mathrm{qq} \to \mathrm{qq}$	$\frac{64}{9}\alpha_s^2\left(\frac{\hat{s}^2+\hat{u}^2}{\hat{t}^2}+\frac{\hat{s}^2+\hat{t}^2}{\hat{u}^2}-\frac{2}{3}\frac{\hat{s}^2}{\hat{u}\hat{t}}\right)$
$\mathrm{q\bar{q}} \to \mathrm{q}'\mathrm{\bar{q}}'$	$\frac{64}{9}\alpha_s^2\left(\frac{\hat{t}^2+\hat{u}^2}{\hat{s}^2}\right)$
$\mathrm{q\bar{q}} \to \mathrm{q\bar{q}}$	$\frac{64}{9}\alpha_s^2\left(\frac{\hat{s}^2+\hat{u}^2}{\hat{t}^2}+\frac{\hat{t}^2+\hat{u}^2}{\hat{s}^2}-\frac{2}{3}\frac{\hat{u}}{\hat{s}\hat{t}}\right)$
$\mathrm{q\bar{q}} \to \mathrm{gg}$	$\frac{128}{3}\alpha_s^2\left(\frac{4}{9}\frac{\hat{t}^2+\hat{u}^2}{\hat{t}\hat{u}}-\frac{\hat{u}^2+\hat{t}^2}{\hat{s}^2}\right)$
$\mathrm{qg} \to \mathrm{qg}$	$16\alpha_s^2\left(\frac{\hat{s}^2+\hat{u}^2}{\hat{t}^2}-\frac{4}{9}\frac{\hat{s}^2+\hat{u}^2}{\hat{s}\hat{u}}\right)$
$\mathrm{gg} \to \mathrm{q\bar{q}}$	$\frac{8}{3}\alpha_s^2\left(\frac{1}{3}\frac{\hat{t}^2+\hat{u}^2}{\hat{t}\hat{u}}-\frac{3}{4}\frac{\hat{t}^2+\hat{u}^2}{\hat{s}^2}\right)$
$\mathrm{gg} \to \mathrm{gg}$	$72\alpha_s^2\left(3+\frac{\hat{t}^2+\hat{u}^2}{\hat{s}^2}+\frac{\hat{s}^2+\hat{u}^2}{\hat{t}^2}+\frac{\hat{s}^2+\hat{t}^2}{\hat{u}^2}\right)$
$\gamma\mathrm{q} \to \mathrm{qg}$	$\frac{128}{3}\alpha_s\alpha\, e_\mathrm{q}^2\left(-\frac{\hat{s}^2+\hat{u}^2}{\hat{s}\hat{u}}\right)$
$\gamma\mathrm{g} \to \mathrm{q\bar{q}}$	$16\alpha_s\alpha\, e_\mathrm{q}^2\left(\frac{\hat{t}^2+\hat{u}^2}{\hat{t}\hat{u}}\right)$
$\gamma\gamma \to \mathrm{q\bar{q}}$	$32\alpha^2\, e_\mathrm{q}^4\left(\frac{\hat{t}^2+\hat{u}^2}{\hat{t}\hat{u}}\right)$

parton cross sections. The rise of the resolved photon processes is similar. However, the resolved processes are predicted to rise more steeply than the direct photon processes. In terms of the transverse parton momentum

$$\hat{p}_t = \frac{\sqrt{\hat{s}}}{2}\sin\hat{\theta}\ , \tag{2.26}$$

the relative contribution of the direct processes to the total cross section increases with $\hat{p}_t$.

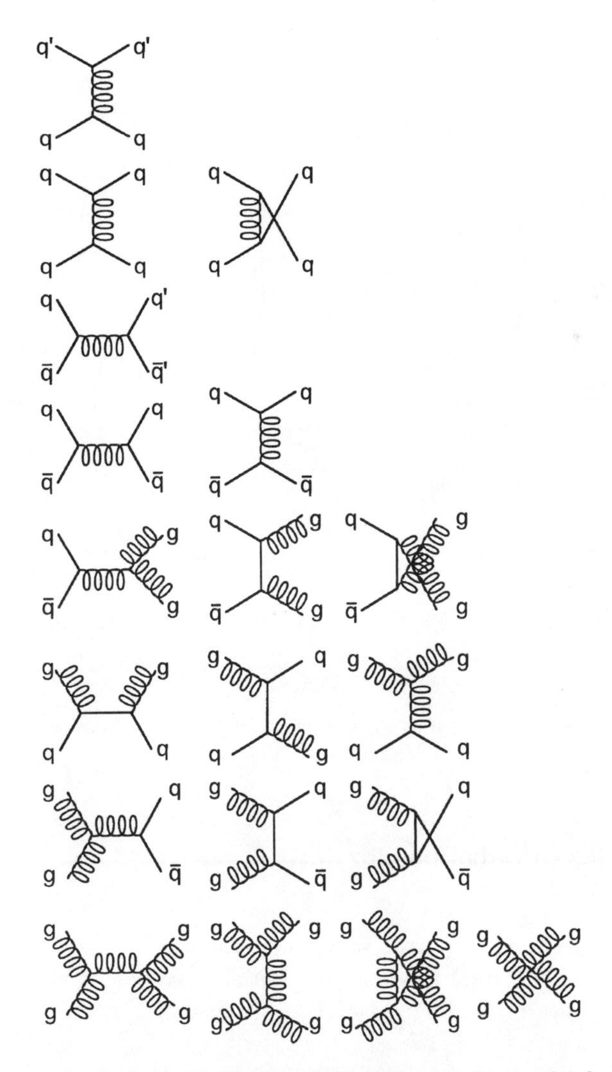

Fig. 2.4. Leading order QCD diagrams for resolved photon–nucleon and double resolved two-photon scattering [84]

The renormalization scale μ_{ren} of the strong coupling constant α_s is set proportional to the transverse momentum $\hat{p}_t$ of the final-state partons. Since α_s has to be small for reliable predictions of perturbative QCD, the parton transverse momentum has to be above some minimum value, usually set to about $\hat{p}_t = 2$ GeV. The small $\hat{p}_t$ region corresponds to the region where the majority of the parton cross sections diverge, that is at small and very large scattering angles (2.26). A lower cut-off in $\hat{p}_t$ therefore simultaneously solves the divergency problem and guarantees small α_s.

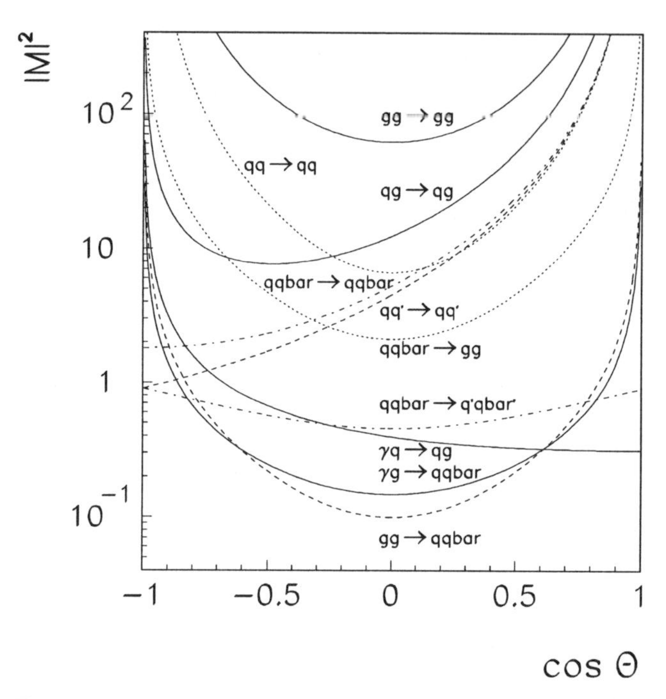

Fig. 2.5. Distributions of the cosine of the parton scattering angle $\hat{\theta}$ in the parton–parton center-of-mass system are shown for different leading order QCD matrix elements

2.2.2 Leading Order QCD Predictions for Jet and Particle Cross Sections

To predict observable cross sections for parton scattering processes in lepton–target interactions involving quasi-real photons, the initial-state parton luminosity and the final-state parton fragmentation to hadrons or jets need to be known.

Parton Luminosity.

1. The electrons radiate $f_{\gamma/\mathrm{e}}(y, Q^2)\mathrm{d}y$ photons with scaled energy $0 < y < 1$ and photon virtuality Q^2. Here the Weizsäcker-Williams approximation (2.11) can be used with high accuracy for sufficiently small $Q^2_{\mathrm{max}} < 1\ \mathrm{GeV}^2$.
2. The parton distribution functions of the photon $f_{i/\gamma}(x_\gamma, \mu_{\mathrm{fac}})$ are given as functions of the scaled parton energy $x_\gamma = E_{\mathrm{parton}}/E_\gamma$, and the factorization scale μ_{fac}. The factorization scale μ_{fac} is usually set equal to the renormalization scale μ_{ren}. If nonperturbative bound state effects in the fluctuations of the photons into $\mathrm{q\overline{q}}$ pairs are negligible, the anomalous photon contribution of (2.18) can be applied. For most applications the hadronic photon contributions are important and parameterizations of

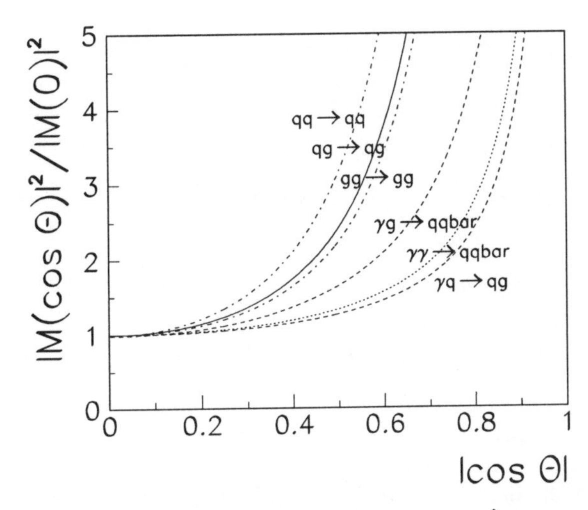

Fig. 2.6. The shapes of the parton angular $\hat{\theta}$ distributions from Fig. 2.5 are compared for different parton scattering processes. For the resolved photon interactions, only those processes are shown that give the dominant contribution to the jet and particle cross sections. Since forward and backward parton scattering cannot be distinguished experimentally, the distributions are shown as a function of the absolute value of $\cos\hat{\theta}$. Matrix elements of resolved γ interactions rise more steeply than those of direct γ processes

$f_{i/\gamma}$, extracted from experimental results, must be used. Their accuracy will be discussed in Sect. 3.

3. The parton distribution functions of the target particle $f_{j/\mathcal{T}}(x_{\mathcal{T}}, \mu_{\text{fac}})$ are required as functions of the scaled parton energy $x_{\mathcal{T}} = E_{\text{parton}}/E_{\mathcal{T}}$, and the factorization scale μ_{fac}.

Fragmentation.

1. The fragmentation of a final-state parton k into a hadron h is denoted by $D_{\text{h}/k}(z, \mu_{\text{frag}})$. Here z is the energy fraction of the hadron relative to the parton energy, $z = E_{\text{h}}/E_k$, and μ_{frag} is the fragmentation scale, which is usually set equal to the hadron transverse momentum $\mu_{\text{frag}} = p_t^{\text{h}}$. The fragmentation functions have been parameterized from results of e^+e^- and $\bar{\text{p}}\text{p}$ experiments. In the case of the quark fragmentation function, the accuracy is of the order of a few percent [15]. The uncertainty in the gluon fragmentation function currently contributes an error of 20% to NLO QCD calculations of γp particle cross sections. This uncertainty will soon be reduced to the precision of the quark fragmentation functions (J. Binnewies and B.A. Kniehl, private communications 1996).
2. The fragmentation of a final-state parton k into a jet is treated by intelligent jet definitions: the jet algorithms are designed to integrate the energy of the parton, which is distributed to several hadrons during the fragmentation process.

To predict the inclusive cross sections of hadron and jet production, the parton cross sections have to be summed over all quarks and gluons, and integrated over all input distributions. The inclusive hadron and jet cross sections are then calculated by

$$\begin{aligned} \frac{\mathrm{d}\sigma^{\mathrm{h}}}{\mathrm{d}t} &= \sum_{ijk} \int \mathrm{d}y \, \mathrm{d}x_\gamma \mathrm{d}x_\tau \mathrm{d}z \; f_{\gamma/\mathrm{e}}(y) \; f_{i/\gamma}(x_\gamma) \; f_{j/\tau}(x_\tau) \\ &\quad \times \frac{\mathrm{d}\hat{\sigma}}{\mathrm{d}\hat{t}}(ij \to kX) \; D_{\mathrm{h}/k}(z) \; , \end{aligned} \tag{2.27}$$

$$\begin{aligned} \frac{\mathrm{d}\sigma^{\mathrm{jets}}}{\mathrm{d}\hat{t}} &= \sum_{ij} \int \mathrm{d}y \, \mathrm{d}x_\gamma \mathrm{d}x_\tau f_{\gamma/\mathrm{e}}(y) \; f_{i/\gamma}(x_\gamma) \; f_{j/\tau}(x_\tau) \\ &\quad \times \frac{\mathrm{d}\hat{\sigma}}{\mathrm{d}\hat{t}}(ij \to \mathrm{jets}) \; , \end{aligned} \tag{2.28}$$

where the dependencies on the different scales μ_{ren}, μ_{fac}, μ_{frag}, and Q^2 have been suppressed to have legible formulae. Apart from these scales, a total of four variables is needed to describe the partonic state. This set of variables can, e.g., be chosen to be $(y, x_\gamma, x_\tau, \hat{t}\,)$. The parton center-of-mass energy $\sqrt{\hat{s}}$, needed to calculate the parton cross section (2.25), can be calculated from the photon and parton fractional energies, and the beam center-of-mass energy $s_{\mathrm{e}\tau}$:

$$\hat{s} = y \; x_\gamma \; x_\tau \; s_{\mathrm{e}\tau} \; . \tag{2.29}$$

Other choices of the set of observables will be discussed in Sect. 2.2.3.

Note that in (2.27, 2.28), the parton cross sections $\mathrm{d}\hat{\sigma}/\mathrm{d}\hat{t}$ are the only components that are 'pure' predictions of QCD theory. The common practice of calling these inclusive cross sections 'QCD predictions' relies on all the input distributions mentioned above, and on the factorization of the different input distributions. Turning the argument around: since QCD has been confirmed by many different experiments, comparisons of data with such calculations potentially give new information on the input distributions f and D.

Contributions of the different processes. The contributions of the individual parton subprocesses to the jet or particle cross sections depend on the kinematical region under consideration. In Fig. 2.7 the relative contributions of the different subprocesses to the inclusive differential jet cross section in photon–proton collisions is shown as a function of the jet transverse energy E_t^{jet}. The names of the subprocesses give in the first letter the parton coming from the photon, in the second letter the parton from the proton. Note that the composition depends on the choice of the parton distributions in the photon and the proton and on the selected range of the jet rapidity.

For the resolved γp interactions, only those processes are shown that give a significant contribution to the total calculation. At small E_t^{jet} these processes dominate the jet cross section. The relative contribution of the direct γp interactions increases with E_t^{jet} and reflects the particular shapes of these matrix elements [Fig. 2.6 and (2.26)].

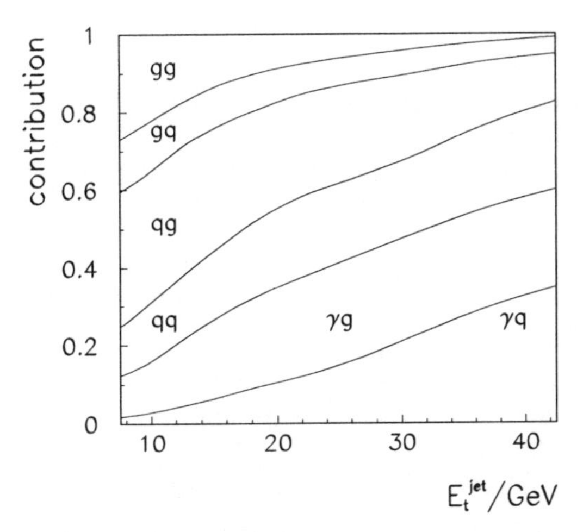

Fig. 2.7. The relative contributions to the jet production in photon–proton collisions are shown as a function of the transverse jet energy.The processes include here the direct γp processes and those resolved γp processes, which give a sizable contribution to the total rate. The jets were found with a cone algorithm using the cone size $R = 0.7$ and jet rapidities between $-2.5 < \eta_{\gamma p}^{\rm jet} < 0.5$, which were determined in the γp center-of-mass system. The contributions were calculated using the PYTHIA event generator with multiple parton interactions (Sect. 2.2.6) together with the GRV-LO parameterizations for the proton and the photon [49, 51]

Single Effective Subprocess and Effective Parton Distribution. Two features of the relevant resolved photon processes allow for a simple approximation of the jet and particle cross sections (2.27, 2.28) [35]: the shapes of the qq, qg, gg matrix elements are very similar (Fig. 2.6) and can be replaced by a single effective subprocess (SES). The relative rates of the matrix elements differ by the ratio of the color factors $C_{\rm A}/C_{\rm F} = 9/4$. The matrix elements are related by $|M_{\rm qq}|^2 : |M_{\rm qg}|^2 : |M_{\rm gg}|^2 = 1 : (9/4) : (9/4)^2$. These color factors are introduced in the definition of so-called effective parton distribution functions

$$\tilde{f}_\gamma \equiv \sum_{\rm n_f} (\rm q_\gamma + \bar{q}_\gamma) + \frac{9}{4} g_\gamma \quad \text{and} \tag{2.30}$$

$$\tilde{f}_\tau \equiv \sum_{\rm n_f} (\rm q_\tau + \bar{q}_\tau) + \frac{9}{4} g_\tau \,, \tag{2.31}$$

where the sums run over the quark flavors. For example, in (2.28), the sum over the resolved processes can be replaced by

$$\sum_{ij} f_{i/\gamma}\, f_{j/\tau}\, \frac{{\rm d}\hat{\sigma}}{{\rm d}\hat{t}}(ij \to \text{jets}) \approx \tilde{f}_\gamma\, \tilde{f}_\tau\, \frac{{\rm d}\hat{\sigma}}{{\rm d}\hat{t}}(\text{SES}) \,. \tag{2.32}$$

This approximation simplifies the extraction of information on the parton content of the photon.

2.2.3 Four Observables to Determine the Parton Kinematics

Convenient observables to describe the final-state partons are their transverse momenta and rapidities in the laboratory system. The transformation from the parton center-of-mass system (PCMS) to the laboratory frame is given by

$$p_t = \hat{p}_t \,, \tag{2.33}$$

$$\eta = \hat{\eta} + \eta_{\rm pcms} \,. \tag{2.34}$$

Here the intrinsic transverse momenta of the partons from the photon and the target particle have been neglected. Since the parton configuration in the PCMS is back-to-back (Fig. 2.3), $\hat{\theta}_1 = \pi - \hat{\theta}_2$, their rapidity sum is $\hat{\eta}_1 + \hat{\eta}_2 = 0$. Using (2.34), the boost of the PCMS with respect to the laboratory system can be calculated by the parton rapidities in the laboratory frame:

$$\eta_{\rm pcms} = \frac{\eta_1 + \eta_2}{2} \equiv \overline{\eta} \,. \tag{2.35}$$

The rapidity difference is Lorentz invariant under boosts along the beam axis:

$$\Delta\eta \equiv \eta_1 - \eta_2 = \hat{\eta}_1 - \hat{\eta}_2 = -2 \ln \tan \frac{\hat{\theta}}{2} \,. \tag{2.36}$$

From this identity and trigonometrical gymnastics, the cosine of the PCMS parton scattering angle results from:

$$\cos\hat{\theta} = \tanh \frac{\Delta\eta}{2} \,. \tag{2.37}$$

At small values of the rapidity difference $|\Delta\eta| < 1$ it is simply $\cos\hat{\theta} \approx \Delta\eta/2$.

The scaled photon energy y was defined in (2.1). The photon four vector $\boldsymbol{\gamma}$ can be calculated from the four vectors of the beam lepton $\boldsymbol{e}$ and the scattered lepton $\boldsymbol{e}'$ via $\boldsymbol{\gamma} = \boldsymbol{e} - \boldsymbol{e}'$ (Fig. 1.4). For quasi-real photons ($Q^2 \approx 0$), y then results from the energies of the beam lepton $E_{\rm e}$ and the scattered lepton $E_{\rm e'}$:

$$y = 1 - \frac{E_{\rm e'}}{E_{\rm e}} \,. \tag{2.38}$$

Alternatively, $\boldsymbol{\gamma}$ can be calculated from the target four vector $\boldsymbol{\tau}$ and all final-state partons with four vectors $\boldsymbol{k}_j$, including the spectator partons, via $\boldsymbol{\gamma} = \sum_j \boldsymbol{k}_j - \boldsymbol{\tau}$. Then the scaled photon energy is given by the transverse momenta $p_{t,j}$ and rapidities η_j of all partons:

$$y = \frac{\sum_j p_{t,j} \, {\rm e}^{-\eta_j}}{2E_{\rm e}} \,. \tag{2.39}$$

The parton fractional energies x_γ and x_τ can be recovered from the two scattered partons with the four-momenta $\boldsymbol{j}_i$ from four-momentum conservation:

$$y x_\gamma \boldsymbol{e} + x_\tau \boldsymbol{\tau} = \boldsymbol{j}_1 + \boldsymbol{j}_2 \,. \tag{2.40}$$

Here $\boldsymbol{e}$ and $\boldsymbol{\tau}$ denote the four-vectors of the lepton and the target particle with the beam energies E_{e} and $E_{\mathcal{T}}$ in the laboratory system. Multiplication of (2.40) with $\boldsymbol{\tau}$ ($\boldsymbol{e}$) leads to expressions for x_γ ($x_{\mathcal{T}}$), e.g., via the transverse momenta p_t and the rapidities η_i of the scattered partons:

$$x_\gamma = \frac{p_t}{2yE_{\mathrm{e}}}\left(\mathrm{e}^{-\eta_1} + \mathrm{e}^{-\eta_2}\right) , \tag{2.41}$$

$$x_{\mathcal{T}} = \frac{p_t}{2E_{\mathcal{T}}}\left(\mathrm{e}^{+\eta_1} + \mathrm{e}^{+\eta_2}\right) . \tag{2.42}$$

So, the choice of the four variables (y, p_t, η_1, η_2) can be used to describe the initial partonic state (2.29, 2.38 or 2.39, 2.41, 2.42). Also the parton final state can be calculated from these observables using (2.25, 2.37): the differential parton cross sections can be rewritten in the form

$$\frac{\mathrm{d}\hat{\sigma}}{\mathrm{d}\hat{p}_t^2} = \frac{\mathrm{d}\hat{\sigma}}{\mathrm{d}\hat{t}}\frac{1}{\cos\hat{\theta}} \tag{2.43}$$

$$= \frac{|M|^2}{16\pi\cosh^4(\Delta\eta)\tanh(\Delta\eta/2)}\;\frac{1}{\hat{p}_t^4} , \tag{2.44}$$

where $\Delta\eta = \eta_1 - \eta_2$.

$(y, p_t, \Delta\eta, \overline{\eta})$ also gives a possible set to describe the partonic state. The fractional energy of the partons is then calculated from:

$$x_\gamma = \frac{p_t}{2yE_{\mathrm{e}}}\;\cosh(\Delta\eta)\;\mathrm{e}^{-\overline{\eta}} , \tag{2.45}$$

$$x_{\mathcal{T}} = \frac{p_t}{2E_{\mathcal{T}}}\;\cosh(\Delta\eta)\;\mathrm{e}^{+\overline{\eta}} . \tag{2.46}$$

The parton center-of-mass energy can be calculated using (2.29, 2.45, 2.46):

$$\hat{s} = \hat{p}_t^2\;\cosh^2(\Delta\eta) . \tag{2.47}$$

In terms of these observables, the different contributions to the jet and particle cross sections can be studied: at fixed rapidity difference, the differential cross sections (2.44) have a dependence like $\hat{p}_t^{-4}$. This strong dependence dominates the logarithmic behavior of the photon structure function $\ln(\hat{p}_t^2/\Lambda_{\mathrm{QCD}}^2)$ (2.19). Therefore, studies of, e.g., transverse jet energy distributions exhibit essentially the features of the parton scattering processes.

The boost $\overline{\eta}$ is related to x_γ (2.45) and $x_{\mathcal{T}}$ (2.46), and therefore to the parton distributions. The parton cross section is independent of the boost. Therefore, studies of, e.g., the inclusive jet rapidity cross section $\mathrm{d}\sigma/\mathrm{d}\eta$, or the di-jet cross section $\mathrm{d}\sigma/\mathrm{d}\overline{\eta}$, give information on the parton distributions of the photon and the target particle.

2.2.4 Next-to-Leading Order QCD Predictions

The leading order (LO) QCD predictions for jet and particle cross sections have a strong dependence on the choice of the renormalization μ_{ren} and the factorization $\mu_{\rm fac}$ scales. Several theory groups have therefore calculated next-to-leading order (NLO) QCD predictions for the inclusive jet and particle cross sections [7, 18, 55, 58, 79, 82]. For the calculation of inclusive jet cross sections, a user-friendly NLO QCD program package has been provided by [101]. NLO predictions have been calculated for di-jet cross sections [78], and for the production of isolated, neutral particles: π° mesons and prompt photons [56].

A comparison of leading and next-to-leading order calculations for inclusive ep jet cross sections is shown in Fig. 2.8 from [100]. The upper histogram shows the differential jet rapidity cross section $d\sigma^2/dp_t^{\rm jet}/d\eta^{\rm jet}$ for $p_t^{\rm jet} = 5$ GeV. The full curve represents a LO QCD calculation. The dashed (dash-dotted) curve is a NLO QCD calculation using the cone jet algorithm [73] with $R = 1$ ($R = 0.7$). The symbols indicate the NLO QCD results using a different cone jet algorithm. The leading and next-to-leading order QCD calculation with $R = 1$ differ by about 15% with only a small dependence on the rapidity of the jet. This means, leading order QCD calculations give a good first estimate of the QCD prediction.

The lower histogram of Fig. 2.8 shows the scale dependence of the jet cross section at $\eta^{\rm jet} = -2$. The variable ξ denotes the choice of the renormalization and factorization scales in multiples of the jet transverse energy, $\mu = \xi p_t^{\rm jet}$. The scale dependence of jet cross section calculations in γp scattering is significantly reduced in NLO QCD compared to the LO calculations.

2.2.5 Multiple Parton Interactions

The QCD parton cross section $d\hat{\sigma}/d\hat{p}_t^2$ has two inherent problems:

1. It diverges towards small transverse momenta $\hat{p}_t$ of the scattered partons (Sect. 2.2.1). This is usually solved by introducing a lower cut-off $\hat{p}_t > \hat{p}_t^{\rm cut}$.
2. The integrated parton cross section rises steeply with the center-of-mass energy. At some energy it exceeds the measured total cross section.

Item 2) is illustrated in Fig. 2.9, which shows the energy dependence of the leading order QCD parton cross section (dashed curve). It was calculated with the PYTHIA event generator (Sect. 2.2.6) using the GRV-LO parameterizations of the parton distributions in the proton and the photon [49, 51] and a transverse momentum cut-off $\hat{p}_t^{\rm cut} = 1.4$ GeV. Also shown is the energy dependence of the total photon–proton cross section, as parameterized by Donnachie and Landshoff [38] (full curve), together with the HERA measurements at $\sqrt{s_{\gamma p}} \approx 200$ GeV (full circle: H1 [67], full triangle: ZEUS [132]). For fixed energy, the problem 2) can be solved by a higher $\hat{p}_t$ cut-off, or by a reduction of the parton density in the photon. But this only shifts the problem towards higher center-of-mass energies.

Another possible solution is to consider the proton and the resolved photon as beams of partons and to allow for $n > 1$ parton interactions in one

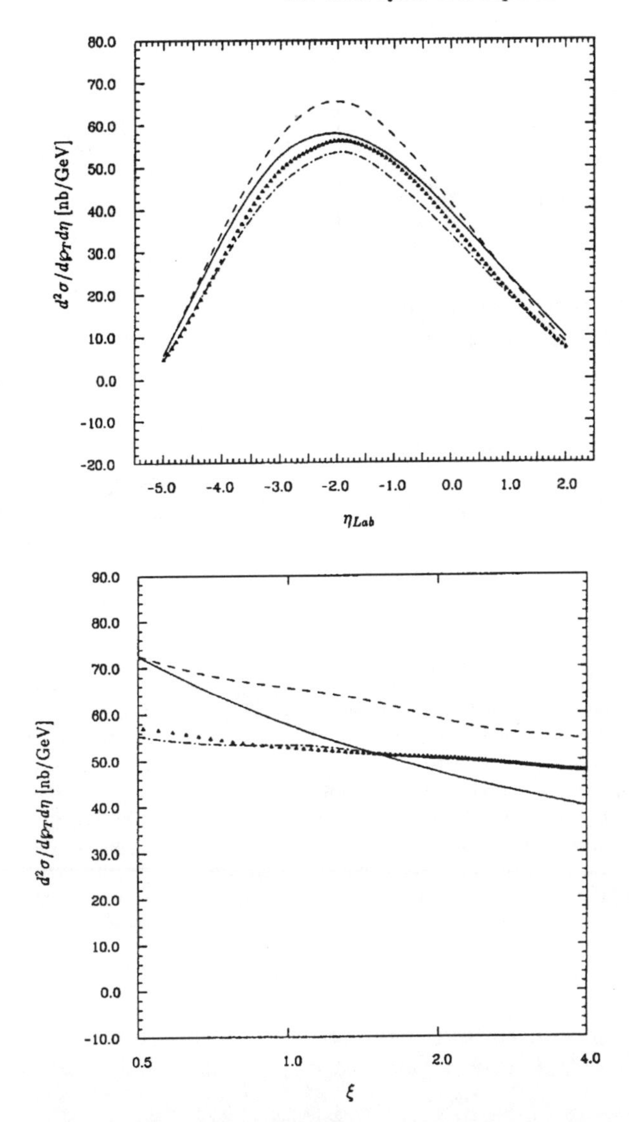

Fig. 2.8. The upper histogram shows different calculations of the differential jet rapidity cross section for $p_t^{\text{jet}} = 5$ GeV. The full curve represents a leading order QCD calculation. The *dashed (dash-dotted) curve* is a next-to-leading order QCD calculation using the cone jet algorithm [73] with $R = 1$ ($R = 0.7$). The *symbols* indicate the result using another cone jet algorithm. The lower histogram shows the dependence of the jet cross section at the jet rapidity $\eta = -2$ on the renormalization and factorization scales in multiples of the jet energy $\mu = \xi\, p_t^{\text{jet}}$

observable γp event [102, 109]. Then the observed inelastic cross section for resolved γp processes is n times smaller than the single parton cross section. Such a mathematical procedure, which produces a varying number of par-

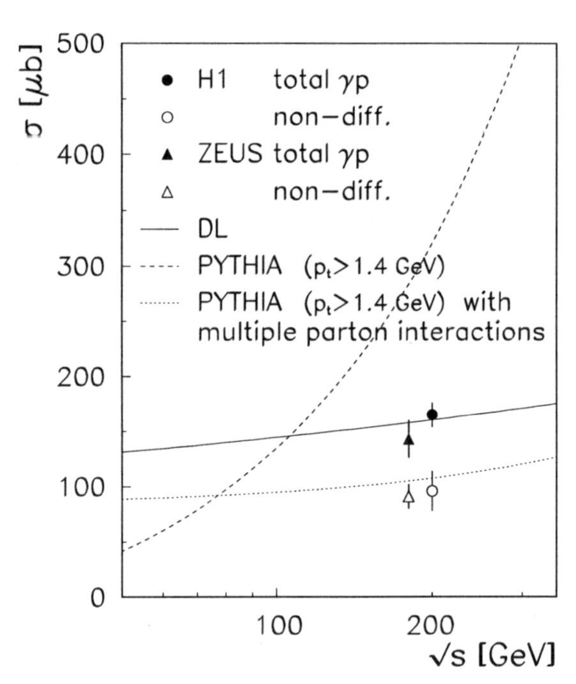

Fig. 2.9. The energy dependence of the leading-order QCD parton cross section (*dashed curve*) for transverse parton momenta above $\hat{p}_t > 1.4$ GeV is shown using the PYTHIA generator (Sect. 2.2.6) together with the GRV-LO parameterizations of the parton distributions in the proton and the photon [49, 51]. The *dotted curve* shows the same calculation allowing for several parton scattering processes in one observable γp event (multiple parton interactions). For comparison, the energy dependence of the total photoproduction cross section is shown as the *full curve* [38]. Also shown are the HERA measurements of the total photoproduction cross section by the *full symbols*, and the nondiffractive cross section by the *open symbols* (*circles*: H1 [67], *triangles*: ZEUS [132])

ton interactions per event and ensures that the total parton cross section stays below the total observed cross section $\hat{\sigma} < \sigma_{\gamma \mathrm{p}}$, is traditionally called 'unitarization'.

The obvious disadvantage of the multiple parton interaction concept is the introduction of additional model assumptions. The advantage of the concept is that the calculated nondiffractive γp cross section stays below the total cross section and describes the HERA measurements (Fig. 2.9 dotted curve, open circle: H1 [67], open triangle: ZEUS [132]). Also, an important problem in the description of the energy flow in jet events is solved: the measured energy flow next to jets is far larger than expected from a single hard parton scattering process and its fragmentation. This energy flow is correctly described by the introduction of multiple parton interactions. The study of the energy flow in jet events turned out to be a key factor for the understanding of the HERA data in terms of the parton content of the photon. It will be discussed in Sect. 4.3.

2.2.6 Event Generators

Quantitative analysis of hard scattering processes requires a detailed understanding of the hadronic final state, which is measured in the detectors, and its connection to the underlying parton dynamics that we want to study. This is the purpose of the QCD event generators.

In Fig. 2.10 a schematic view of a QCD event generator for electron–proton scattering is shown. In the central part of the program are the parton scattering processes (2.25) together with the matrix elements of Table 2.1. The partons from the photon and the target particle coming into the hard subprocess are chosen via the parton distribution functions $f_{i/\gamma}$ and $f_{j/\mathrm{p}}$. Since incoming and outgoing partons can radiate other partons, so-called initial-state and final-state parton showers simulate higher order QCD processes. All partons, including those of the beam remnants, fragment into hadrons. The programs predict the exclusive hadronic final states of hard parton collisions.

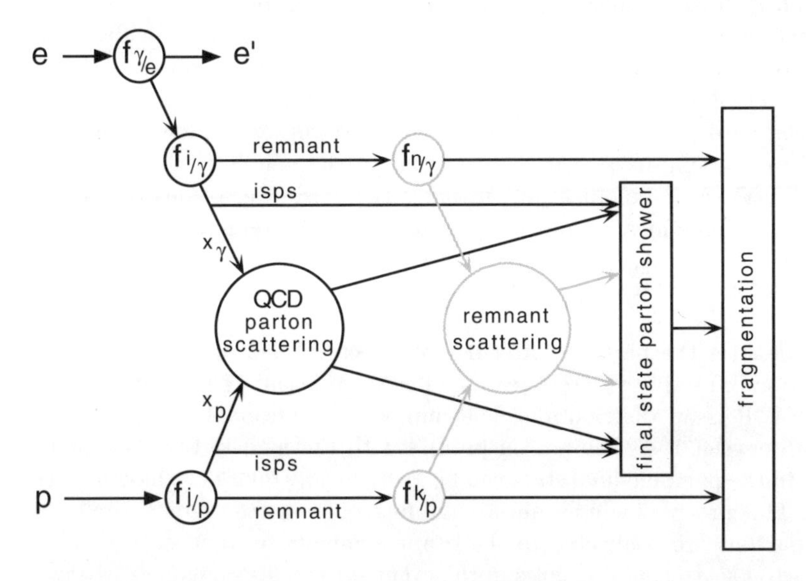

Fig. 2.10. The schematic view of a generator for hard scattering processes in electron–proton collisions is shown: the core is given by the QCD parton cross section (2.25). The input parton distribution functions are labeled $f_{i/\gamma}$ and $f_{j/\mathrm{p}}$. The incoming and outgoing partons can radiate other partons (isps=initial-state parton shower, final-state parton shower). Together with the beam spectator partons (remnant) they fragment into hadrons (fragmentation). Optionally, interactions between the two beam remnants can be generated in addition to the primary hard parton scattering process

The generators, described below, have the option of allowing interactions between the beam remnants in addition to the hard parton scattering process. The treatment of the remnant interactions is different in each of the generators, which allows the study of the model dependencies of additional interactions.

PYTHIA. The PYTHIA 5.7 event generator is usually run in the mode of photon–proton interactions [110], together with a generator for quasi-real photons. PYTHIA is based on the leading order (LO) QCD matrix elements. It includes initial- and final-state parton radiation effects, which are calculated in the leading logarithmic approximation. In most comparisons with data, the strong coupling constant α_s is calculated in first order QCD using $\Lambda_{\text{QCD}} = 200$ MeV with 4 flavors, and the renormalization and factorization scales are both set to the transverse momentum $\hat{p}_t$ of the partons emerging from the hard interaction. Since the leading order QCD calculation of a single hard parton scattering is divergent for processes with small transverse momenta $\hat{p}_t$, a lower cut-off has to be applied. The lowest cut-off values chosen for comparisons with HERA data are around $\hat{p}_t^{\text{cut}} \approx 2$ GeV. For the hadronization process the LUND string fragmentation scheme is used (JETSET [108]).

Within PYTHIA, multiple parton interactions may be generated in addition to the primary parton–parton scattering [102]. These are calculated as leading order QCD processes between partons from the photon and proton remnants. The PYTHIA multiple parton interaction model extends the concept of the hard perturbative QCD parton scattering to the low transverse momentum, or semi-hard scattering region. In the simplest version of the model, the transverse momentum cut-off of the hard interactions is lowered to $\hat{p}_t^{\text{mia}} < \hat{p}_t^{\text{cut}}$. The mean number $\langle n \rangle$ of (semi-) hard interactions is given by

$$\langle n \rangle = \frac{\hat{\sigma}\left(\hat{p}_t > \hat{p}_t^{\text{mia}}\right)}{\sigma_{\text{nondiff.}}} , \tag{2.48}$$

where $\hat{\sigma}$ denotes the parton scattering cross section, and $\sigma_{\text{nondiff.}}$ is the observable nondiffractive γp cross section. The fluctuations of n are calculated based on a Poisson distribution. The number of additional interactions is typically between 1–4. The parton process with the highest transverse momentum in the partonic final state can be given by any quark or gluon matrix element. This process includes initial- and final-state parton radiation effects and its partons are connected to the beam remnants by strings. The additional parton scattering processes in the event are calculated as perturbative gluon–gluon scattering processes.

PHOJET. The PHOJET 1.0 event generator was designed to simulate in a consistent way all components that contribute to the total photoproduction cross section [43, 44]. It is based on the two-component dual parton model [28]. The implementation of the PHOJET generator is similar to the Monte Carlo event generator DTUJET [5, 20], which simulates multi-particle production in high-energy hadron collisions. This latter generator was originally intended for the description of soft hadronic interactions and was then extended to hard scattering processes. In contrast to PYTHIA, PHOJET incorporates very detailed simulations of both multiple soft and hard parton interactions on the basis of a unitarization scheme [27]. The soft hadronic processes are described by the soft, 'supercritical' Pomeron [19, 45]. These processes are simulated by a two-string ansatz, which allows for initial transverse momenta of the partons at the ends of the strings. The hard processes are calculated using the leading order QCD matrix elements. Final-state parton radiation effects are simulated using the JETSET 7.4 program [108]. Hard initial-state parton radiation has recently been included in the program version 1.05. The lower momentum cut-off for hard parton interactions was chosen to be $\hat{p}_t^{\mathrm{cut}} = 3$ GeV. Owing to the unitarization scheme, small variations of this cut-off parameter do not have a large influence on the results of this generator. The model parameters that describe the soft part of the γp interactions have been tuned using results from proton–anti-proton collisions and low-energy photoproduction cross section measurements. For the fragmentation, the LUND string concept is applied (JETSET 7.4 [108]).

HERWIG. The HERWIG 5.8 ep generator is also based on leading-order QCD calculations [88]. This program was designed to have as much input from perturbative QCD as possible, in order to minimize the free parameters. HERWIG includes a parton shower model, which allows for interference effects between the initial- and final-state showers (color coherence) [85, 86]. The renormalization and factorization scales are set according to the transverse momentum of the scattered partons with a lower cut-off $\hat{p}_t^{\mathrm{cut}} \geq 2\,\mathrm{GeV}$. A cluster model is used to simulate the hadronization effects [87, 125].

HERWIG also has the option of allowing for additional interactions between the beam remnants. These interactions are called the *soft underlying event* and are parameterizations of experimental results on 'soft' hadron-hadron collisions. So far, a tuning of the strength and frequency parameters has not been successful. Recently, a model for multiple parton interactions has also been developed for HERWIG, which gives a much more promising approach to describing the data [25].

Table 2.2. Comparison of different leading order QCD Generators: all models include the leading order QCD matrix elements (hard interactions), parton radiation effects to simulate higher order effects, and a simulation of the fragmentation phase. Some generators offer, in addition, options for interactions between the beam remnants, a modeling of the transition region between soft and hard processes, or the soft interactions themselves

Generator		soft ia	hard ia QCD	unitarization	beam remnant ia	parton showers	hadronization
PYTHIA	1		•			initial+final state	string
	2		•	•	hard parton ia		
	3	•					
PHOJET		•	•	•	soft, hard parton ia	final state	string
HERWIG	1		•			initial+final state plus interference	cluster
	2		•		soft und.event		
	3		•	•	hard parton ia		

Summary

1. QCD predicts the cross sections of parton–parton scattering as functions of the parton center-of-mass energy, and the parton scattering angle.
2. For predictions of jet and particle cross sections, the parton luminosities and fragmentation functions are needed as input. Comparisons of measurements with leading and next-to-leading order QCD predictions provide, on the one hand tests of perturbative QCD, and on the other hand, they give information on the parton distributions and fragmentation effects.
3. Four observables are needed to calculate the final-state parton kinematics in leading order, e.g., the scaled photon energy y, the transverse parton momentum p_t, and the parton rapidities η_1, η_2. These variables are conveniently related to observables in di-jet events.
4. For the prediction of the exclusive final state in photon–nucleon and photon–photon interactions event generators have been provided that are based on leading order QCD calculations.
5. In resolved photon–nucleon interactions, or double resolved photon–photon scattering, the photon and the nucleon can be considered as beams of partons such that several parton pairs may collide in one observable event (multiple parton interactions).

3. Two-γ Physics: Deep Inelastic Lepton–Photon Scattering

Around 1980, the electron–positron collider experiments started to study photon–photon interactions. Here we are interested in the $\gamma\gamma$ processes, shown in Fig. 1.3, where one photon is quasi-real ($Q^2 \approx 0$ GeV^2) and the second photon is highly virtual ($Q^2 \gg 1$ GeV^2). This process can be regarded as deep inelastic scattering (DIS) from a real photon target.

Measurement of the double-differential cross section as a function of Q^2 and the parton fractional energy x_γ gives the structure function F_2^γ of the quasi-real photon:

$$\frac{\mathrm{d}^2\sigma}{\mathrm{d}x_\gamma\,\mathrm{d}Q^2} = \frac{4\pi\alpha^2}{x_\gamma\,Q^4}\left(1 - y + \frac{y^2}{2}\right) F_2^\gamma(x_\gamma, Q^2)\,. \tag{3.1}$$

Here α is the fine structure constant and y denotes the scaled photon energy. y is related to Q^2 and x_γ via $Q^2 = x_\gamma y s_{\mathrm{ee}}$ where $\sqrt{s_{\mathrm{ee}}}$ is the $\mathrm{e}^+\mathrm{e}^-$ center-of-mass energy.

In the following two sections we will discuss the experimental results on F_2^γ and summarize the parton distribution functions, which were extracted from the measurements.

3.1 Measurements of the Photon Structure Function F_2^γ

A compilation of all F_2^γ measurements, available in 1991, is shown in Fig. 3.1 (from [123]). The photon virtuality varies in the range $0.7 < Q^2 < 100$ GeV^2, covering parton fractional energies up to $x_\gamma = 0.9$. The accuracy of the measurements varies between 10–50%. The data appear essentially flat in x_γ, but show a significant rise with increasing Q^2.

Recently, smaller x_γ values became available at high Q^2 from experiments at the TRISTAN and LEP colliders. In Fig. 3.2 F_2^γ measurements of the TOPAZ, OPAL, and DELPHI experiments are compared at similar average $\langle Q^2\rangle \approx 15$ GeV^2 [37, 114, 124]. The agreement between the two LEP measurements is quite good. They show a decrease towards small x_γ. The TOPAZ measurement on the other hand shows the tendency to increase towards small x_γ. The data points were lowered here by the QPM charm contribution calculated in [114] in order to compare them with the LEP measurements.

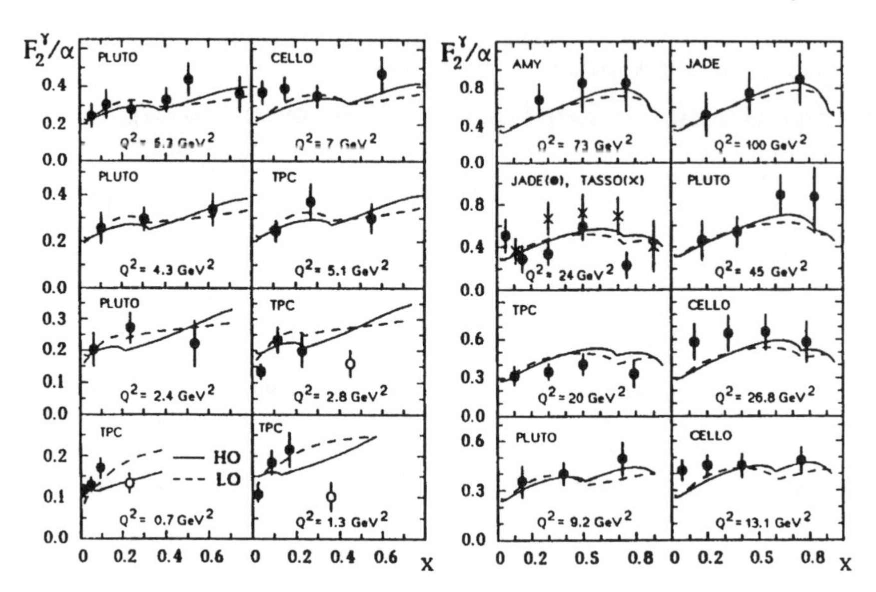

Fig. 3.1. Measurements of the structure function F_2^γ of the quasi-real photon are shown from experiments before 1991. The data (*full symbols*) are drawn as a function of the parton energy fraction x_γ and at different average photon virtuality Q^2. The data are compared with the leading order (*dashed*) and next-to-leading order (*full*) parameterizations by GRV [51]. The *full points* were used to fit the parameter κ of this parton distribution function (from [123])

The data are compared to four parameterizations of the parton distributions of the photon, calculated at $\langle Q^2 \rangle = 12\ \text{GeV}^2$ [37]. The different curves demonstrate the uncertainties in the extrapolations towards the small x_γ region. For new parameterizations of the parton distributions of the photon at small x_γ, the difference between the experimental results needs to be clarified.

The anomalous photon component can be tested by the remarkable parton energy distribution and the scale dependence, predicted in (2.19). In Fig. 3.3, the Q^2 dependence of the data is studied at large parton fractional energies between $0.3 < x_\gamma < 0.8$. The structure function F_2^γ rises with increasing Q^2 at a rate that is compatible with a linear dependence on $\log Q^2$. In Fig. 3.4 the PLUTO measurement of F_2^γ is shown for $\langle Q^2 \rangle = 5.3\ \text{GeV}^2$ as a function of x_γ [14]. The data are clearly distinct from a falling distribution, as would be expected from a hadronic type structure function (long-dashed curve labeled 'HAD'). Both the x_γ distribution and the scale dependence confirm the QCD prediction on the anomalous component of the photon. Note that even at large x_γ the hadronic contribution to F_2^γ is sizable at $Q^2 = 5\ \text{GeV}^2$. Since this part cannot be calculated by perturbative QCD, a precise determination of $\alpha_s \propto \left(\ln\left(Q^2/\Lambda^2_{\text{QCD}}\right)\right)^{-1}$ from the measured scale dependence is difficult.

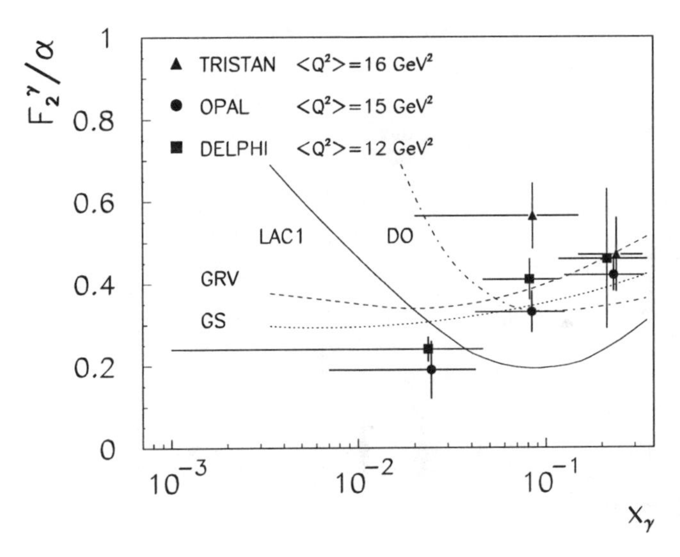

Fig. 3.2. The photon structure function F_2^γ was measured at average photon virtuality $\langle Q^2 \rangle \approx 15\ \mathrm{GeV}^2$ in the interval of the parton fractional energy $0.001 \leq x_\gamma \leq 0.35$ (TRISTAN: TOPAZ *triangles* [114], LEP: OPAL *circles* [124], DELPHI *squares* [37]). Only statistical errors are shown in the figure. The curves represent the photon structure function as obtained with different parameterizations of the parton distributions [37]: *full*=LAC1 [1], *dashed*=GRV [51], *dotted*=GS [53], *dash-dotted*=DO [41]

Summary

1. Two-photon experiments have established the anomalous photon component which was predicted by QCD theory.
2. The measurements of the structure function F_2^γ constrain the quark distribution of the photon in the range of parton fractional energies $0.05 < x_\gamma < 0.9$ to the level of $10 - 50\%$. At small $x_\gamma < 0.1$ and $Q^2 \approx 15\ \mathrm{GeV}^2$, differences between recent measurements at TRISTAN and LEP need to be clarified.
3. The anomalous photon component cannot be separated in a unique way from the hadronic part of the photon structure function. Therefore, for meaningful prediction of, e.g., jet production rates in $\gamma\gamma$ or γp scattering, the parton distributions need to include both the anomalous and the vector meson components of the photon.

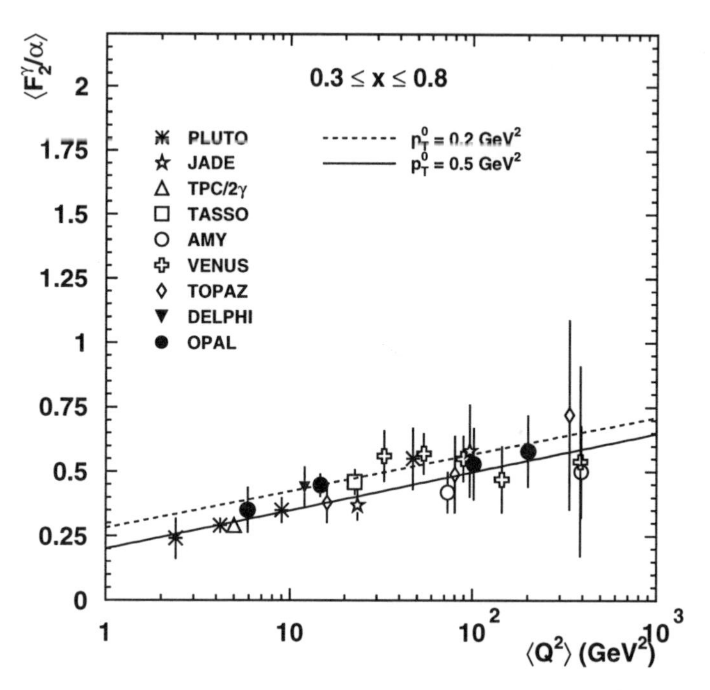

Fig. 3.3. Measurements of the photon structure function F_2^γ are shown as a function of the scale Q^2 for a fixed interval of parton fractional energies $0.3 < x_\gamma < 0.8$. The data (*symbols*) are compatible with a linear dependence of F_2^γ on $\log Q^2$, as predicted by QCD theory. The *curves* are the calculations of the FKP model [47, 48] using different cut-off values for the model parameter p_T° (compilation by B. Kennedy, 1996)

3.2 Parameterizations of the Parton Distributions in the Photon

The parton distributions of the photon can be extracted from the measurements of the photon structure function using (2.14) and the evolution equations (2.16, 2.17). For convenient application, parameterizations of the parton distributions were introduced, which contain the parton momentum distributions at some scale $Q_\circ^2$ and their Q^2 evolutions. Almost all of them are used somewhere by experiments. Here a short summary from the user's point of view is given, for recent theoretical reviews refer to [111, 123]. Meanwhile more than a dozen parameterization sets exist with most of them available via the PDFLIB [96]. The parameterizations can be subdivided into three categories:

Pointlike/anomalous component of real photons.

1. The first parameterizations were constructed by Duke and Owens (DO) who made an ansatz for the pointlike coupling of the photon and used

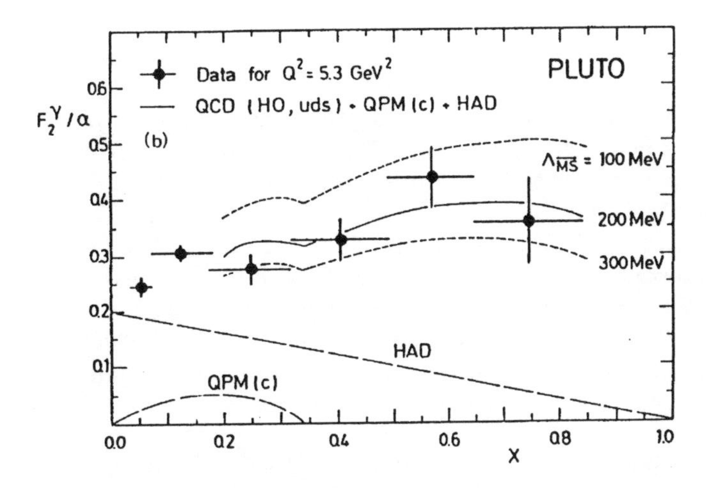

Fig. 3.4. The photon structure function F_2^γ is shown as a function of the parton fractional energy x_γ at the average scale $\langle Q^2 \rangle = 5.3$ GeV^2 (*full circles*, PETRA: PLUTO Collab.) The measured F_2^γ increases with x_γ, which is clearly distinct from a hadronic structure function, and arises from the pointlike coupling of the photon to quarks. The *long-dashed curve HAD* indicates the structure function of a vector meson. The *curve QPM* reflects the charm contribution as calculated in the quark parton model. The *full curve* shows the summed QPM and HAD contributions together with a higher order QCD calculation for the light quarks. The *short dashed curves* indicate the sensitivity of the calculation to the QCD parameter Λ_{QCD} (reprinted from [14] with kind permission from Elsevier Science - NL, Sara Burgerhartstraat 25, 1055 KV Amsterdam, The Netherlands)

an asymptotic solution for the QCD evolution equations [41]. The parton distributions of the vector meson part of the photon had to be added 'by hand', e.g., from the TPC/2γ measurements [117].

2. Field, Kapusta and Paggioli (FKP) provided parameterizations of the pointlike (i.e., anomalous) photon component including higher order QCD effects. The parameterizations contain the Q^2 evolution at large x_γ [47, 48, 76].

Complete parameterizations of the real photon.

1. The first successful attempt to parameterize the anomalous and vector meson parts of the photon structure together was made by Drees and Grassie (DG) [39]. Using an ansatz for the parton distributions at $Q^2 = 1$ GeV^2 together with the leading order inhomogeneous evolution equations, they fitted first data points from deep inelastic eγ scattering experiments [12].
2. This ansatz was carried further by Levy, Abramowicz and Charchula (LAC) [1] using more than 50 data points. They demonstrated that the gluon distribution in the photon was not constrained by the DIS eγ results. Accordingly, they provided three different test sets of leading order

parameterizations of photonic parton distributions with different gluon contents.
3. In a similar approach by Watanabe, Hagiwara, Izubuchi and Tanaka, the charm contribution to F_2^γ was treated in an elaborate way: 6 WHIT parameterizations were obtained, each with different gluon distributions in the photon [70].
4. Gordon and Storrow (GS) provided several leading and next-to-leading order parameterizations where, for example, they used a gluon input at $Q^2 = 5.3$ GeV^2 equivalent to that of a vector meson [53]. They allowed for a κ-factor for the overall uncertainty of this ansatz, and fitted κ and the light quark masses to the DIS eγ data. Recently they provided new parameterizations that make use of the jet results from $\gamma\gamma$ and γp scattering [54].
5. Glück, Reya and Vogt (GRV) provided leading and next-to-leading order parameterizations, which are constructed at very low $Q_\mathrm{o}^2(\mathrm{LO}) = 0.25$ GeV^2 [51]. Here the valence quark distributions in the photon have the same shape as in the pion structure function [50]. The gluon content was set proportional to the valence quark content. The difference between the π and the γ distributions results from the first term in the evolution equation (2.16). They fit only 1 parameter, essentially the same κ-factor as GS, to the DIS eγ data (Fig. 3.1).
6. Aurenche, Fontannaz and Guillet (AFG) provided a next-to-leading parameterization with a more elaborate ansatz for the vector meson input at low Q_o^2 [6].

Complete parameterizations for real and virtual photons.

1. Glück, Reya and Stratmann (GRS) extended the very low Q_o^2 ansatz for real photons and studied the photon at virtualities ≈ 1 GeV^2 [52]. Nonperturbative contributions to the photon structure are predicted to be sizable here as well.
2. Schuler and Sjöstrand (SAS) fitted the shapes of their parameterization to real photon data and constrained the normalization via the VDM. In addition to the parameterizations for real photons, they provide them also for virtual photons based on a dispersion relation in the photon mass [104, 105].

A comparison of the main features of the different parameterizations is shown in Table 3.1. Examples for different parameterizations of the parton distributions in the photon are shown in Fig. 3.5 from [111]. The leading and next-to-leading order quark parameterizations agree in the range $0.1 < x_\gamma < 0.9$ to the level of 30%. At large $x_\gamma > 0.9$, the parameterizations show large differences; most striking is the steep increase of the GRV-NLO and AFG-NLO distributions. At small $x_\gamma < 0.1$, the large differences between the extrapolation of different leading order quark parameterizations were shown in Fig. 3.2. The gluon distribution is essentially unconstrained by the DIS eγ data, as

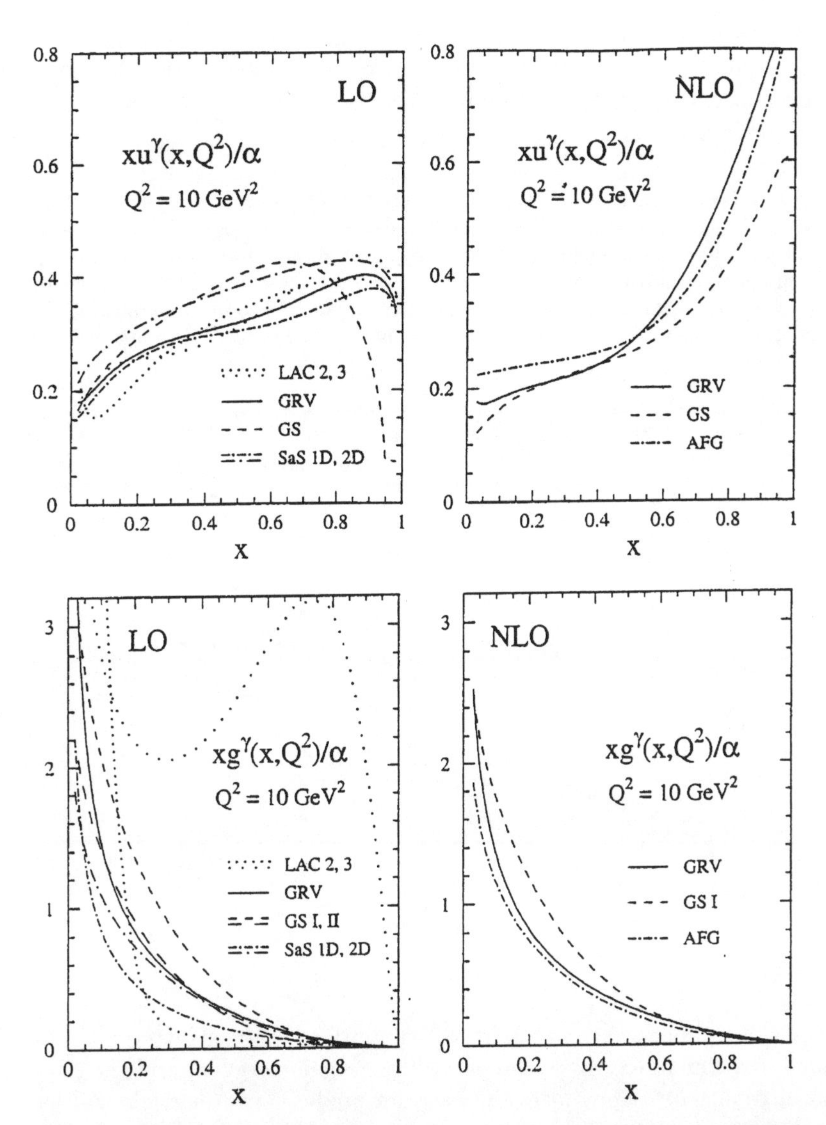

Fig. 3.5. Different leading order (**left**) and next-to-leading order (**right**) parameterizations for the parton distributions in the photon are compared as functions of the fractional energy x_γ. The **upper** two histograms show the u-quark distribution, which is known to the level of 30% in the range $0.1 \leq x_\gamma \leq 0.9$, but has large uncertainties at very small and very large x_γ. The **lower** two histograms refer to the gluon in the photon, which has essentially no constraints from the results of deep inelastic $e\gamma$ scattering experiments. The similarity of the different next-to-leading order gluon parameterizations arises from a similar modeling of the gluon in the photon, but not from photon data (reprinted from [111] with kind permission from Institute of Physics Publishing, Techno House, Redcliffe way, Bristol BS1 6NX, United Kingdom)

demonstrated by the different LAC parameterizations in Fig. 3.5.The next-to-leading order gluon parameterizations look very similar. This however is not a consequence of the data, but a result of similar theoretical input, based on the VDM.

Table 3.1. Parameterizations of parton distributions in real and virtual photons in leading and next-to-leading order QCD. They contain the pointlike (i.e., anomalous) component of the photon, or both the pointlike and hadronic (i.e., VDM) components of the photon. The parameterizations were constructed at a fixed scale $Q_\circ^2$. Higher values of the scale are calculated with the QCD evolution equations. To constrain the gluon in the photon, some parameterizations use information from vector mesons

authors	year	real γ	virtual γ	LO	NLO	pointlike/ anomalous	hadr./ VDM	$Q_\circ^2$ (GeV^2)
DO	82	•		•		•		
FKP	86	•			•	•		
DG	85	•		•		•	•	1
LAC	91	•		•		•	•	1/4
WHIT	95	•		•		•	•	4
GS	92/96	•		•	•	•	•	5.3/3
GRV	92	•		•	•	•	•	0.25/0.3
AFG	94	•			•	•	•	0.25
GRS	94	•	•	•	•	•	•	0.25/0.3
SAS	95	•	•	•		•	•	0.6/2

Summary

1. From the F_2^γ measurements, parton distributions were extracted and parameterized as functions of the parton fractional energy x_γ and the photon virtuality Q^2. They form the basis for predictions of particle and jet production, e.g., in collisions of two quasi-real photons at KEK and LEP, or in hard photon–proton scattering at HERA.
2. The gluon distribution of the photon is essentially unconstrained by the results of the deep inelastic $e\gamma$ scattering experiments.

4. Photon–Proton Interactions at HERA

4.1 Introduction

The lepton–hadron collider HERA allows the study of a rich variety of different physics processes. In this introduction to photoproduction physics at HERA, we want to:

1. list the relevant photoproduction processes,
2. define the final-state variables,
3. identify the kinematic region where scattering of partons from the photon and the proton can be observed, and
4. explain the main features of the machine and the detectors.

4.1.1 Photoproduction Processes

In Fig. 4.1 an overview of the photoproduction processes at HERA is given. Three types of processes can be distinguished: elastic, diffractive processes and nondiffractive processes.

They can be further separated into soft and hard processes according to the transverse energy produced in the hadronic final state. The soft processes include elastic vector meson production, photon dissociation, proton dissociation and double dissociation. Soft nondiffractive processes dominate the total cross section (Fig. 2.9, [67, 132]). Such processes can be understood as peripheral collisions of a vector-meson-like photon and the proton.

Hard scattering processes that lead to particles with high transverse momenta and jets have been observed in nondiffractive processes [60, 130], and recently also in diffractive scattering [63, 134]. Here direct and resolved photon processes are predicted by QCD.

In this review, we are interested in the hard scattering processes leading to particles with high transverse momenta and jets, in order to study the photon and its partonic interactions. We concentrate on the study of nondiffractive hard scattering, since here the parton luminosities of the proton provide well-defined input distributions for the QCD calculations.

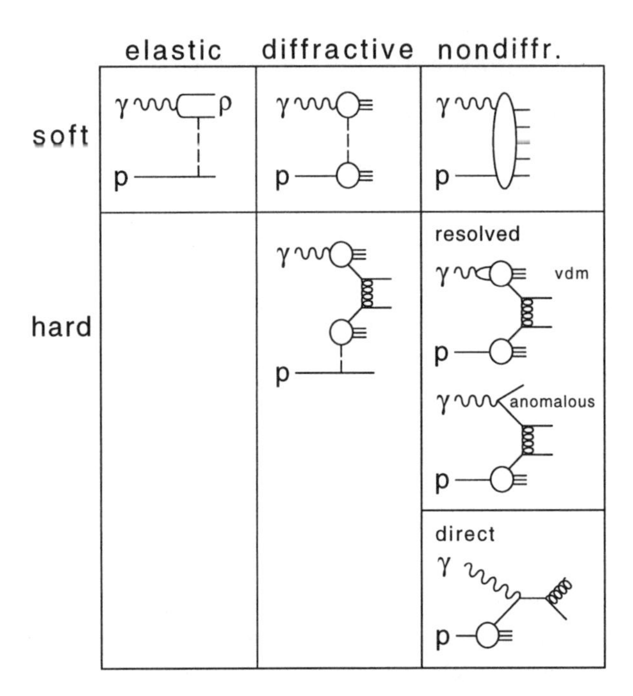

Fig. 4.1. Photoproduction processes are labeled '*soft*' and '*hard*', according to the transverse energy E_t produced in the hadronic final state. The soft processes include elastic contributions, e.g., ϱ production, single and double dissociation, and nondiffractive interactions, i.e., peripheral photon–proton scattering processes. In the nondiffractive *hard* processes, the quarks and gluons of the proton are used to probe the quasi-real photons. In *resolved* photon–proton interactions, the photon fluctuates into a $q\overline{q}$ pair before the hard scattering process occurs. The pair can form a vector meson state (*VDM*) or propagate without forming a hadronic bound state (*anomalous*). In *direct* photon–proton interactions, the bare photon couples to a quark of the proton. In diffractive events, *hard* scattering processes have recently been established. These events are also expected to have direct and resolved γp processes (this separation is not shown in the figure)

4.1.2 Event Kinematics

In this section we define the final-state variables and draw the kinematical region where the photoproduction events appear. We indicate the jet configurations resulting from small and large parton energies and sketch the detector coverage in the photon–proton center-of-mass system.

Variables to Describe the Final-State Particles . To describe the final-state particles, three variables are conventionally used, which are drawn in Fig. 4.2:

1. the transverse momentum p_t is measured with respect to the beam axis,
2. φ describes the azimuthal angle around the beam pipe, and

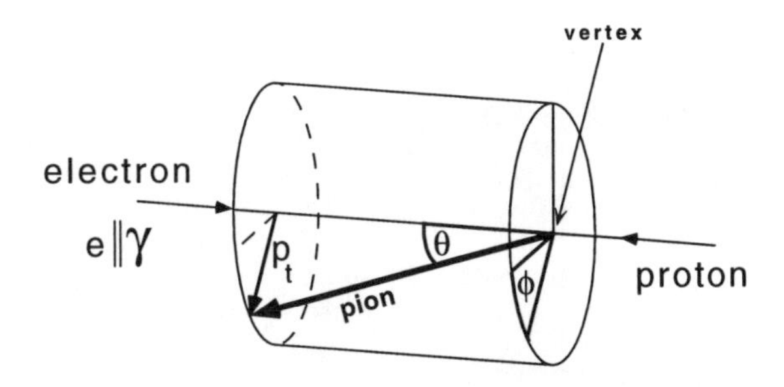

Fig. 4.2. Variables for the final-state particles: the transverse momentum p_t is measured with respect to the beam axis. φ denotes the azimuthal angle around the beam axis. The polar angle θ is measured relative to the proton direction. Instead of θ the pseudo-rapidity $\eta = -\ln\tan(\theta/2)$ is frequently used

3. θ is the polar angle with respect to the proton direction. It is frequently replaced by the laboratory pseudo-rapidity, defined as $\eta = -\ln\tan(\theta/2)$. We will also make use of the pseudo-rapidity in the photon–proton center-of-mass system, which we denote $\eta_{\gamma\mathrm{p}}$.

Photon Virtuality, Center-of-Mass Energy, and Transverse Energy. The available phase space at HERA is sketched in Fig. 4.3 as functions of the photon virtuality Q^2, the photon–proton center-of-mass energy $\sqrt{s_{\gamma\mathrm{p}}}$, and the summed transverse energy E_t of the final-state hadrons.

We measure continuously from $Q^2 \approx 0$ GeV^2 (photoproduction) to large values of Q^2 (deep inelastic lepton–proton scattering) with a transition region around $Q^2 \approx 1$ GeV^2. The transition region is very interesting and is presently being studied by several theory groups and by the HERA experiments. For the purpose of this review, we concentrate on quasi-real photons, which are best selected in the kinematical region below $Q^2 < 10^{-2}$ GeV^2 and between $150 < \sqrt{s_{\gamma\mathrm{p}}} < 250$ GeV by the detection of the scattered lepton in the electron detector of the luminosity system (dashed window: 'tagged electron').

In order to increase statistics, events are also analyzed in which the electron disappears undetected at small scattering angles with respect to the beam direction with average virtuality $Q^2 = 10^{-3}$ GeV^2 [134], and a maximum virtuality of $Q^2 = 4$ GeV^2 ('untagged events'). The photon energy is derived from the scattered electron energy (2.38) in the case of electron-tagged events and in the untagged case from the final-state hadrons (2.39).

For the study of hard scattering processes, events with sufficiently high transverse energy E_t are required (E_t at least 10 GeV).

Parton Kinematics. The parton fractional energies, x_γ and x_p, can be reconstructed in two-jet events using (2.41, 2.42). In Fig. 4.4, three jet configurations are shown. At fixed parton transverse momentum and fixed photon energy, the configurations correspond to:

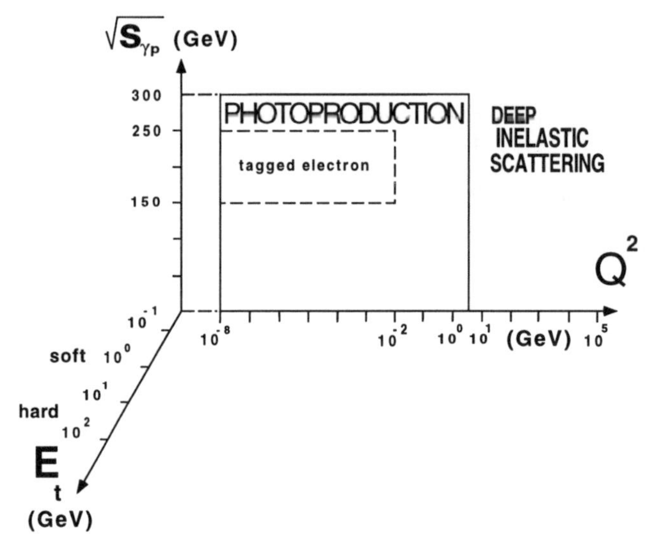

Fig. 4.3. Relevant scales for photon–proton scattering at HERA: the negative squared four-momentum of the exchanged photon is $Q^2 \equiv -\gamma^2$ (Fig. 1.4). Above $Q^2 \approx 4$ GeV^2 the interacting leptons are scattered into the main detectors (Figs. 4.7, 4.8). These events are attributed to deep inelastic ep scattering processes. The region below $Q^2 \approx 1$ GeV^2 is termed the 'photoproduction' region, i.e., the scattering of quasi-real photons with protons. $\sqrt{s_{\gamma\mathrm{p}}}$ is the photon–proton center-of-mass energy. The dashed line indicates the kinematic window where the scattered lepton can be found in the small-angle electron detector of the luminosity system. E_t denotes the summed transverse energy produced in the hadronic final state. The low E_t region, labeled 'soft', refers to peripheral scattering processes between the proton and vector meson states of the photon. High E_t corresponds to the region where hard parton scattering processes between the photon and the proton are expected

(a) *low* x_γ and *high* x_p: both jets are found at large rapidities η,
(b) *high* x_γ and *high* x_p: one jet is at positive, one jet at negative rapidities,
(c) *high* x_γ and *low* x_p: both jets are at rapidity $\eta \approx 0$.

Transverse Energy as a Function of Rapidity. The boost of the photon–proton center-of-mass system (CMS) with respect to the laboratory system is about $\Delta\eta = 2$. In Fig. 4.5 a schematic view of the transverse energy distribution is shown as a function of the photon–proton center-of-mass rapidity $\eta_{\gamma\mathrm{p}}$. The additional axes indicate the laboratory variables η and θ. The proton remnant particles appear as the enhancement at positive rapidities. The region within ± 1 units around the γp rapidity $\eta_{\gamma\mathrm{p}} = 0$ we call the 'mid-rapidity' region. The negative (positive) rapidity region $\eta_{\gamma\mathrm{p}} < -1$ ($\eta_{\gamma\mathrm{p}} > 1$) corresponds to the 'photon fragmentation region' ('proton fragmentation region'). The scattered electron is indicated at negative rapidities (not to scale).

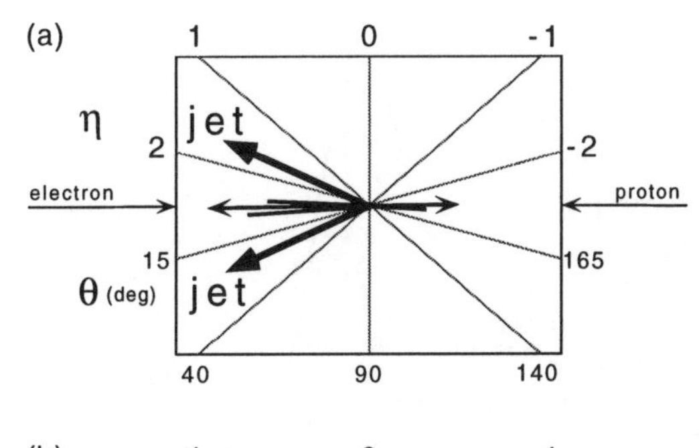

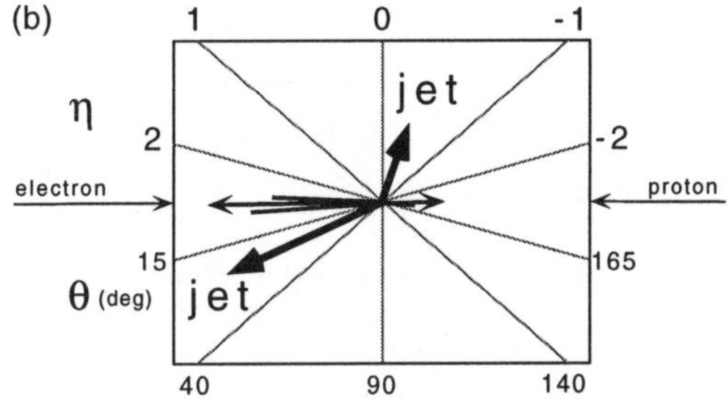

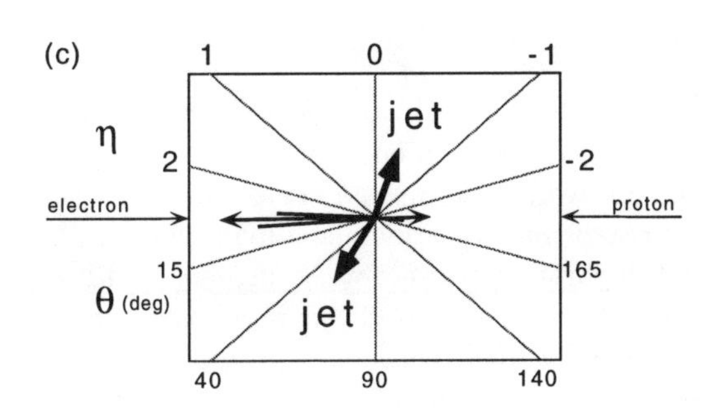

Fig. 4.4. Different jet configurations correspond to different regions of the parton kinematics. (**a**) In events with low x_γ and high x_p, both jets are typically found at large rapidities η. (**b**) At high x_γ and high x_p, a typical configuration is that one jet is at positive, one jet at negative rapidities. (**c**) At high x_γ and low x_p, both jets are at rapidity $\eta \approx 0$

The geometric acceptance of the H1 and ZEUS detectors and their electron detectors is indicated by the dashed lines. The main detectors cover a large fraction of the photon fragmentation region, the central photon–proton collision region, and only a small part of the proton fragmentation region.

4.1.3 The HERA Accelerator

A new generation photon facility came into being in 1992 when the first lepton-proton collider 'Hadron-Elektron-Ring-Anlage' (HERA) was set up at DESY (Hamburg, Germany). The center-of-mass energies are one order of magnitude above the previous photon experiments and reach up to $\sqrt{s_{\rm ep}} \approx$ 300 GeV. The separate storage rings for the protons and the electrons have a circumference of 6.3 km and provide 4 interaction regions. The beam energies of the proton $E_p = 820$ GeV and the electron $E_e = 26.7$ GeV are largely

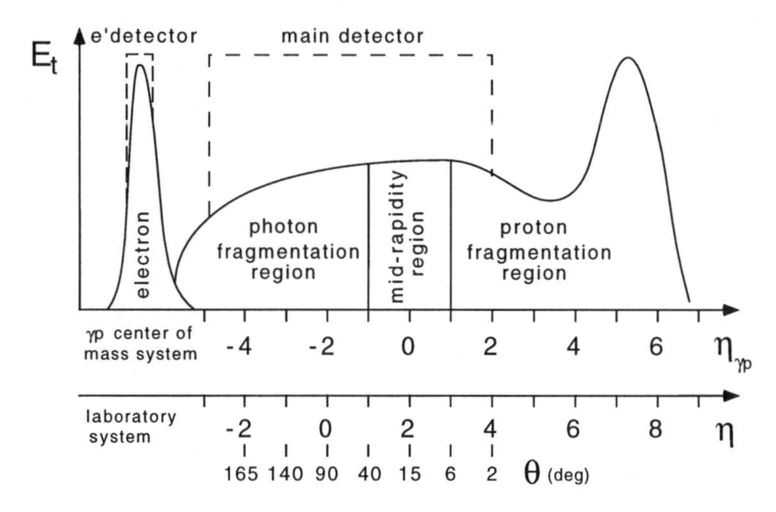

Fig. 4.5. A schematic view of the transverse energy flow is drawn as a function of the photon–proton center-of-mass rapidity $\eta_{\gamma p}$. Also shown are the corresponding observables in the laboratory frame, rapidity η and polar angle θ, and the position of the main detector. The electron distribution is indicated on the left side (not to scale)

asymmetric, satisfying the need for high center-of-mass energies at minimal synchrotron radiation losses. In 1994, the electron energy was raised slightly to $E_e = 27.6$ GeV, and later positrons were filled instead of electrons with the advantage of more stable running conditions. In principle, a maximum of 210 bunches of electrons and protons, separated in time by 96 ns, can be filled. For the purpose of controlling the background in the experiments, about 10 electron (proton) bunches are usually filled without a proton (electron) bunch partner. Variation of the interaction points along the beam direction is determined by the proton bunch lengths, and shows an approximately Gaussian distribution with a width of $\sigma = 11$ cm. A proton fill typically lasts one day, a positron fill around 7 h.

The luminosity has increased each year since the start-up of HERA and has reached $L = 7.7 \times 10^{30}$ cm^{-2}s^{-1}, which corresponds to about half the design luminosity. The integrated luminosity produced for the years 1992–96 is shown in Fig. 4.6. It provides sufficient statistics for the study of photoproduction physics. In the left histogram, the luminosity produced by the HERA machine is shown; the right histogram represents the luminosity used by the H1 experiment for data taking.

4.1.4 The H1 and ZEUS Experiments

The asymmetric beam energies imply a boost of the final-state particles into the proton direction: e.g., a pion, which was scattered at 90° in the photon–proton center-of-mass system, is at $\theta \approx 15°$ in the laboratory system

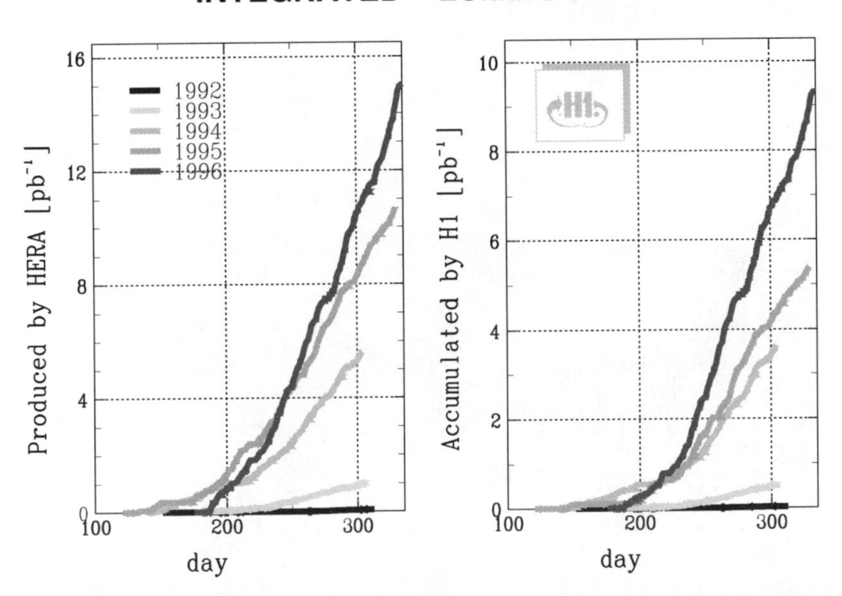

Fig. 4.6. Integrated luminosity versus the day of the year: in the **left** histogram the integrated luminosity produced by the HERA machine is shown. The luminosity has increased each year since the start-up of HERA in 1992, and has now reached up to 50% of the design luminosity 1.5×10^{31} cm^{-2}s^{-1}. About half of the produced luminosity is used for data taking (**right** histogram from H1 Collaboration)

(Fig. 4.5). The energy of our pion in the laboratory system is approximately a factor four above its center-of-mass energy. The detector design, e.g., the interaction length covered by the calorimeters, was adapted according to these conditions.

The H1 and ZEUS collaborations designed multi-purpose detectors (Figs. 4.7 and 4.8). Elaborate descriptions exist in the literature [69, 131]. Here, only a short description of those parts that are relevant to this review is given, with special emphasis on the differences between the detectors, which led to partly complementary achievements in the photoproduction sector.

Both experiments have several tracking devices that detect charged particles over the full azimuth and in the polar angular region between $5° \leq \theta \leq 172°$. The trackers are located inside a strong magnetic field with 1.15 (1.4) Tesla in the H1 (ZEUS) experiment. Published results on charged particle production in photoproduction processes rely on the central cylindrical drift chambers, which detect tracks in the interval $20° \leq \theta \leq 150°$. The precision in the transverse momentum measurements of the particles is $\sigma_{p_t}/p_t = 0.0075\, p_t \oplus 0.01$ $(0.005\, p_t \oplus 0.016)$, with p_t in GeV for the H1 (ZEUS) detector.

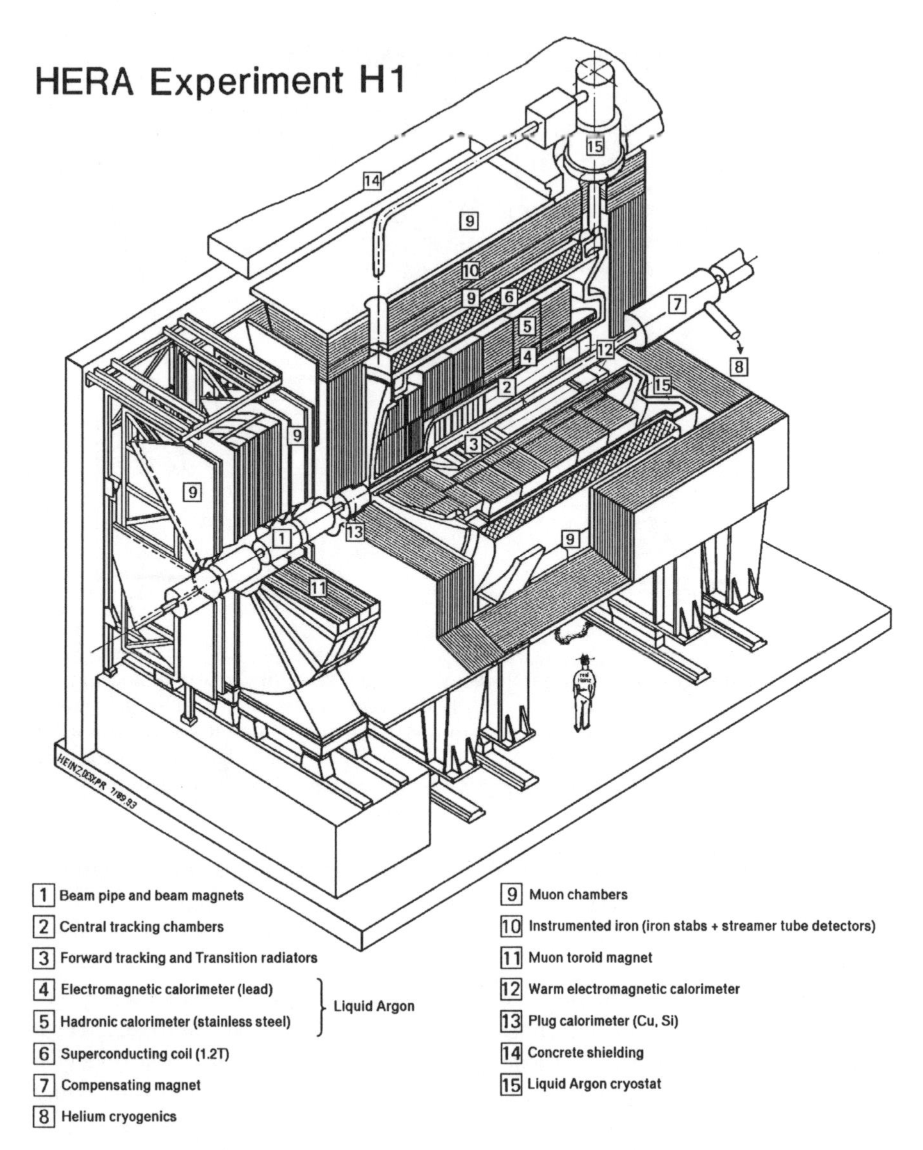

Fig. 4.7. The H1 detector at HERA consists of tracking detectors inside the field of a super-conducting magnet. They are surrounded by elaborate calorimetry (liquid argon, iron, lead scintillator). A separate muon spectrometer is indicated on the left side

The tracking detectors are surrounded by calorimeters with electromagnetic and hadronic sections. The ZEUS collaboration built a uranium-scintillator calorimeter with photomultiplier read-out. The thickness of the uranium and scintillator layers was optimized for the hadronic energy resolution: electrons and pions with the same energy give the same energy re-

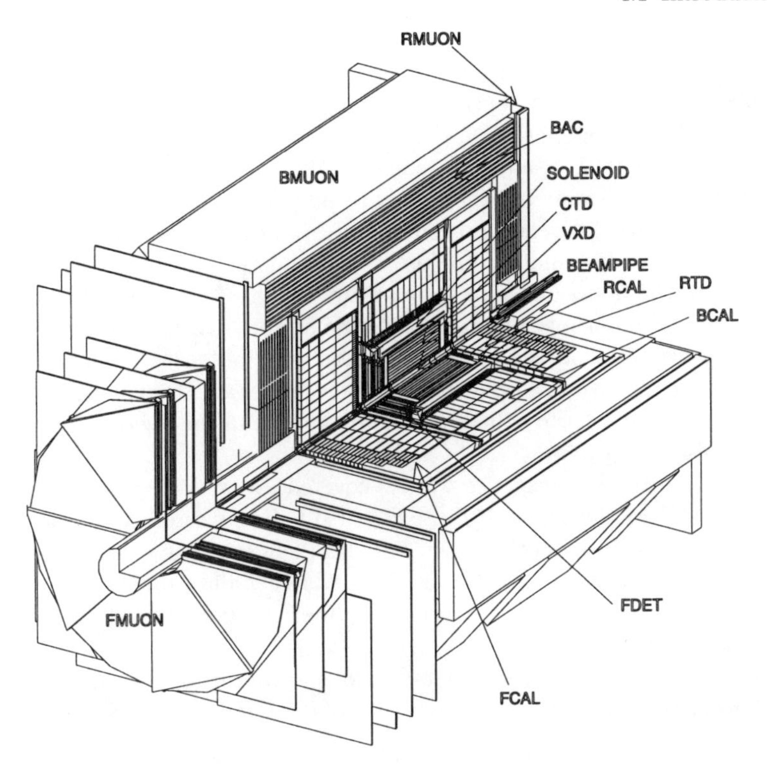

Fig. 4.8. The ZEUS detector at HERA consists of tracking detectors (VXD, CTD, FDET, RDET) inside the field of a warm magnet. They are surrounded by a uranium-scintillator calorimeter with photomultiplier read-out (FCAL, BCAL, RCAL), and an iron calorimeter (BAC). Muon detection is provided (BMUON, RMUON, FMUON)

sponse in the calorimeter ('compensation'). The calorimeter covers the polar angular interval $2.6° \leq \theta \leq 176.1°$. It is mechanically separated into three parts: rear (electron direction), barrel, and forward calorimeter (proton direction). The size of the stacks in the electromagnetic section is $5 \times 20\ \text{cm}^2$ ($10 \times 10\ \text{cm}^2$ for the rear calorimeter) with a depth of 25 radiation lengths. The hadronic stacks are $20 \times 20\ \text{cm}^2$ in size. The total absorption corresponds to 7–4 interaction length, decreasing with increasing θ. The energy resolution for electrons is $\sigma_E/E = 0.18/\sqrt{E} \oplus 1\%$ with E in GeV, and for pions $\sigma_E/E = 0.35/\sqrt{E} \oplus 1\%$. For photon–proton collisions, this calorimeter is especially well suited to measure the hadronic energy depositions over a wide range in the photon fragmentation region of the γp center-of-mass system with a single calorimeter type.

The corresponding calorimeter system of H1 consists of two main parts: a liquid argon calorimeter (LAr) and a lead scintillator calorimeter (1992–94: BEMC, since 1995: SPACAL). The LAr calorimeter covers the polar angle between $4° \leq \theta \leq 153°$. The electromagnetic section consists of lead absorbers and is 20–30 radiation lengths deep. The hadronic section was built with steel absorbers. The total depth of the calorimeter varies between 8–4.5 interaction lengths, depending on θ. The energy resolution for electrons is $\sigma_E/E = 0.12/\sqrt{E} \oplus 0.01\%$, and for pions $\sigma_E/E = 0.5/\sqrt{E} \oplus 0.02\%$. In contrast to the ZEUS calorimeter, the energy responses of pions and electrons with equal energies are different, and require offline corrections according to the shower shape and the energy. In this H1 calorimeter, not only was the depth adapted to the boost of the photon–proton center-of-mass system, but the cell sizes were also optimized according to the asymmetric beam energies: fine granularity was built in the proton direction with cell sizes as small as $3 \times 3\ \mathrm{cm}^2$, increasing with θ. In photoproduction processes, this calorimeter is therefore well suited to study the hadronic final state in the region around the γp center-of-mass system.

The electron direction was previously covered by a lead scintillator calorimeter BEMC in the range $151° \leq \theta \leq 177°$. It consisted of an electromagnetic section with a depth of 22.5 radiation lengths. The energy resolution for electrons was $\sigma_E/E = 0.1/\sqrt{E}$. Since the 1995 data-taking period, the BEMC has been replaced by a scintillating fiber calorimeter with photomultiplier read-out (SPACAL).

Both experiments have verified the hadronic energy scale of the calorimeters to the level of 4–5%, based on comparisons of data and simulated events:

- The ZEUS experiment compared the summed transverse momenta of tracks pointing into jet cones and the calorimetric jet energy measurements in the central detector region.
- They studied the forward calorimeter energy scale with transverse energy balance in di-jet photoproduction events, where one jet points into the central region, the other jet into the proton-forward region.
- The H1 experiment compared the transverse energy balance between the scattered electron and the hadronic final state in deep inelastic scattering events.

The calorimeters of both experiments are surrounded by iron tail catchers, which record energy leakage out of the calorimeters and provide muon identification. In the proton direction, separate devices were built to detect 1) muons, e.g., for charm analysis, 2) protons to study diffractive scattering, and 3) neutrons for measurement of the pion structure function.

The luminosity available to the experiments is measured via the Bethe-Heitler process ep $\to$ epγ which is predicted by QED theory. Both experiments have two electromagnetic calorimeters in the electron direction with photomultiplier read-out: a photon detector for scattering angles below 0.5 mrad at a distance of about 100 m away from the experiments, and an electron detector for scattering angles below 6 mrad at around 30 m (Fig. 4.9). The detectors consist of TlCl/TlBr crystals in the case of H1 with a resolution of $\sigma_E/E = 0.1/\sqrt{E}$. The ZEUS experiment has lead-scintillator calorimeters with similar resolution. The luminosity can be calculated from a simultaneous e and γ tag. Since the background rate in the photon detectors has been found to be negligible, the luminosity is actually calculated from the events in the photon tagger, where no complicated magnetic beam optics have to be simulated to determine the acceptance of the detector.

In the case of γp interactions, the electron detectors of the luminosity systems are especially important. About 40% of the events with scaled photon energies between $0.2 < y < 0.8$ and photon virtuality below $Q^2 < 0.02\ \mathrm{GeV}^2$ have the scattered electron tagged in these detectors: the electron tag ensures that the interacting photon is quasi-real.

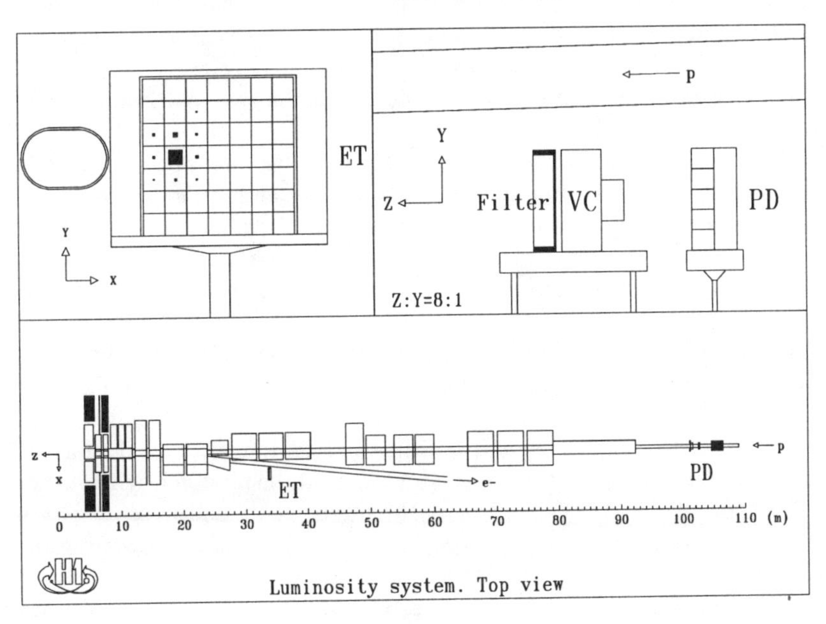

Fig. 4.9. The H1 Luminosity system consists of an electron detector (ET) and a photon detector (PD). The **lower** picture shows a top view of the beam line relative to the interaction point in the direction of the scattered electron. The **upper left** picture shows the front view of the electron tagger together with a typical electron signal. The **upper right** picture shows the side view of the photon detector with a lead filter and a veto counter in front as a protection against synchrotron radiation

In the ZEUS experiment, the main triggers used for photoproduction analysis are calorimeter triggers where towers are formed with low energy thresholds of 0.4 (1.0) GeV in the electromagnetic (hadronic) sections. They are used in coincidence with an energy deposit in the electron tagger above $E \geq 5$ GeV, or as a stand-alone trigger with a summed energy requirement of $E \geq 10$ GeV.

In the case of H1, the main trigger is an energy deposit in the electron detector above $E \geq 4$ GeV in coincidence with at least one track seen by the central drift-chamber trigger with a transverse momentum above

$p_t \geq 0.45$ GeV. For triggering of untagged hard photoproduction events, the electron detector requirement is replaced by an energy deposit in the H1 liquid argon calorimeter. In addition to a summed transverse energy of $E_\mathrm{t} \geq 6$ GeV, a single energy cluster of $E \geq 2$ GeV is required, which is validated by a local coincidence with a track originating from the interaction vertex.

Summary

1. At the HERA ep collider, quasi-real photons are emitted by the leptons. Quarks and gluons of the photon are probed by the partons of the proton at γp center-of-mass energies up to $\sqrt{s_{\gamma\mathrm{p}}} = 300$ GeV.
2. Photon–proton interactions are studied with the two multi-purpose detectors H1 and ZEUS:
   a) The collisions of quasi-real photons with protons are identified by leptons, which scatter through small angles and are measured in the electron detectors of the luminosity system ('tagged electron events'). Events where the electron disappears undetected at small scattering angles are also assigned to photoproduction processes ('untagged events').
   b) For studies of the hadronic final state, the detectors were optimized differently. The H1 liquid argon calorimeter has fine granularity in the central γp collision region. This enables detailed studies of the transverse energy flow and multiple parton interactions. The ZEUS uranium calorimeter covers a large range of the photon fragmentation region. This enables studies of the direct photon processes and the photon remnant jet. Both experiments study the parton distributions of the photon by measurements of the differential jet rates.
   c) For studies of charged particle production with high transverse momentum both experiments have tracking detectors inside a magnetic field, which cover approximately 3 rapidity units of the photon fragmentation region. These devices enable studies of the parton content in the photon through differential cross section measurements.

4.2 Verification of QCD Predictions in γp Scattering

In this section, we want to test the following QCD predictions for hard photoproduction at HERA:

1. Hard scattering between the partons of the photon and the proton should be seen. Such γp interactions lead to particles with large transverse momenta and jet production.
2. Direct and resolved γp interactions should exist. In the leading order approximation of perturbative QCD, they can be distinguished by photon

spectator particles, which should appear only in the case of resolved photon scattering.
3. Direct and resolved γp interactions should have different distributions of the parton scattering angle. This can be studied by distributions of the jet rapidities in di-jet events.
4. Higher-order QCD effects should appear: in di-jet events they cause an imbalance between the jet transverse energies. Multi-jet production should be observed.

4.2.1 Hard Scattering Processes: Particle Production

The first question is: do we see hard parton–parton scattering processes at all? One possible means of answering this is the analysis of inclusive particle spectra. This has been successfully used, e.g., in proton–proton scattering at the ISR in 1972 [75]. While studying π° production for $\theta \approx 90^\circ$, they found two components in the transverse momentum spectrum (Fig. 4.10):

1. At small transverse momenta $p_t < 1$ GeV, the p_t spectrum was found to decrease exponentially as $\exp(-b\,p_t)$ with $b = 6$. This region corresponds to soft hadronic collisions.
2. At large transverse momenta $p_t > 2$ GeV, they measured an excess of high p_t particles, which is described by a power law, in agreement with expectations from hard scattering processes.

Single particle spectra have also been measured by the H1 and ZEUS experiments in γp interactions [62, 137]. The ZEUS experiment measured charged hadron production in the central tracking detector, which covers a large rapidity interval in the photon fragmentation region ($-3.4 < \eta_{\gamma\mathrm{p}} < -0.8$). Hadrons with transverse momenta between $0.3 < p_t < 8$ GeV were selected. The scattered electron was required to be found at small angles in the electron detector of the luminosity system, which ensured the collision of quasi-real photons with protons at photon virtualities below $Q^2 < 0.02$ GeV^2. The corresponding average γp center-of-mass energy was $\sqrt{s_{\gamma\mathrm{p}}} = 180$ GeV. Two types of event sample were selected from the 1993 data-taking period with an integrated luminosity of 0.4 pb^{-1}:

1. Diffractive events were selected by requiring a rapidity region close to the proton direction that was free of hadrons (a so-called 'rapidity gap'). The invariant mass M_X of all hadrons in the final state was $M_\mathrm{X} = 5$ GeV on average, a sample which essentially contains soft scattering processes. The shape of the resulting transverse momentum spectrum of the hadrons is shown in Fig. 4.11 as open circles. The spectrum was fitted with an exponential function (dotted curve).
2. In a nondiffractive event sample (no requirement of a rapidity gap) two components were found in the transverse momentum spectrum (full circles):

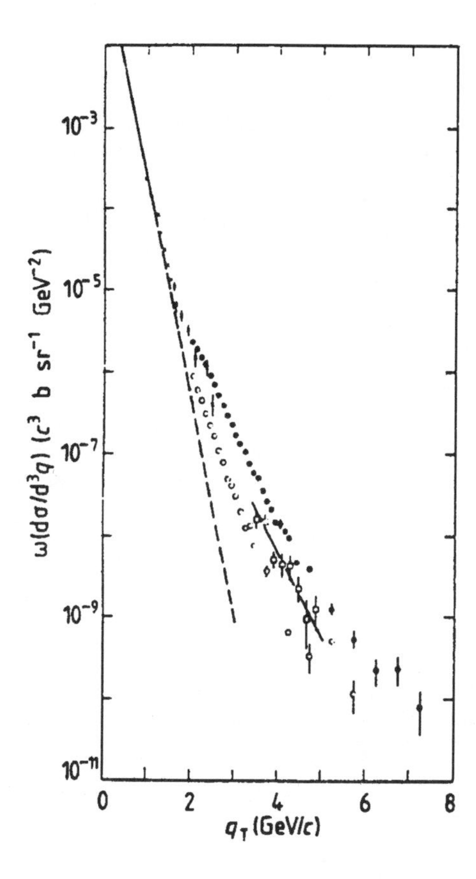

Fig. 4.10. The differential cross section of π° production at 90° in pp collisions at the ISR is shown as a function of the π^0 transverse momentum. The symbols refer to different center-of-mass energies: *open circles*=31 GeV, *open squares*=44 GeV, *full circles/crosses*=53 GeV. The *dashed line* is the extrapolation of $\exp(-6\,p_t)$, which describes the data at $p_t < 1$ GeV (reprinted from the XVI International Conference on High Energy Physics, Batavia, IL (1972) [75] with kind permission from Fermi National Accelerator Laboratory)

a) At small transverse momenta $p_t < 1.2$ GeV, the data can be described by an exponential fit, as in the case of pp collisions. This region corresponds to soft collisions between the photon and the proton.
b) At large transverse momenta $p_t > 1.2$ GeV, the data follow a power law shape, as expected for partonic scattering processes between the photon and the proton. A fit to the functional form $A(1 + p_t/B)^{-n}$ gave $n = 7.25 \pm 0.04$.

Also shown in Fig. 4.11 is a next-to-leading order (NLO) QCD calculation as the full curve, which gives a good description of the shape of the measured cross section, but slightly overshoots the rate of the data [79]. The dash-dotted curve represents the shape of a power law fit to data from $\overline{\mathrm{p}}\mathrm{p}$ collisions at similar center-of-mass energies $\sqrt{s_{\overline{\mathrm{p}}\mathrm{p}}} = 200$ GeV [121]. These data fall significantly more steeply ($n = 12.14 \pm 0.39$) than the photoproduction data and indicate the different parton distributions of photons and protons, and the presence of direct photon processes.

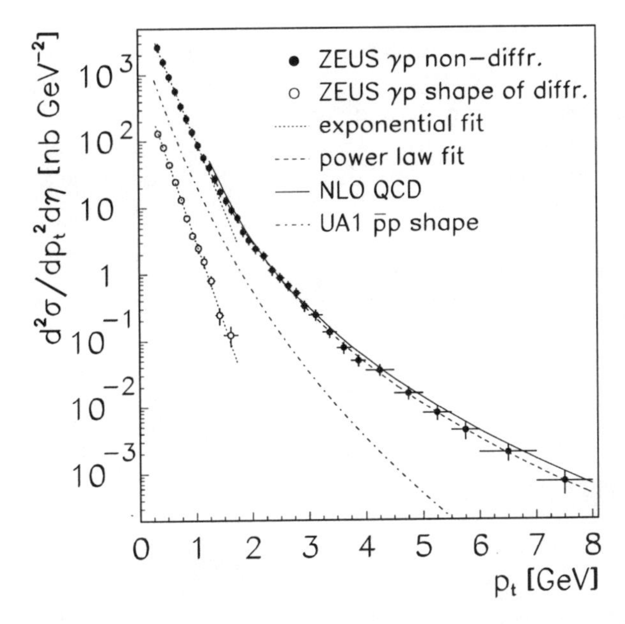

Fig. 4.11. The inclusive differential particle cross section $d\sigma/dp_t$ is shown from ZEUS data as *full circles* [137]. The shape of a diffractive event selection is shown by the *open circles*. The *dotted* (*dashed*) curves represent exponential (power law) fits to the data. The *full curve* is a NLO QCD calculation by [79]. The *dash-dotted curve* is a power law fit to UA1 data from $\bar{p}p$ collisions at similar center-of-mass energies [121]

Summary

1. Parton collisions in photoproduction events at HERA are observed. At peak luminosities, the rate of such events is large (~ 0.1 Hz).
2. Quantitative descriptions of the data are obtained from NLO QCD hard scattering processes, with parton fluxes derived from standard photon- and proton-parton parameterizations, and fragmentation functions determined with e^+e^- and $\bar{p}p$ data. These results confirm QCD theory and show that the quark distribution functions, extracted from deep inelastic $e\gamma$ scattering experiments are applicable to the photoproduction data. Later we will discuss how the HERA results improve the knowledge of the parton distributions in the photon (Sect. 4.4).

4.2.2 Hard Scattering Processes: Jet Production

The second proof of the existence of hard scattering processes in photon–proton collisions was the observation of jets with large transverse energy. Their existence is beyond any doubt: the jets can be seen 'by the eye' on event displays. In Fig. 4.12 a photoproduction event, observed with the H1 detector, is displayed demonstrating that the individual jets are, in fact, strongly focused energy depositions. This event is a candidate for a direct photon process, since no energy deposition of a photon remnant is observed in the direction of the scattered electron. A candidate event for jet production from a resolved photon process is shown from ZEUS data in Fig. 4.13: here a third jet is observed in the electron direction (right), which shows the particles of the photon remnant.

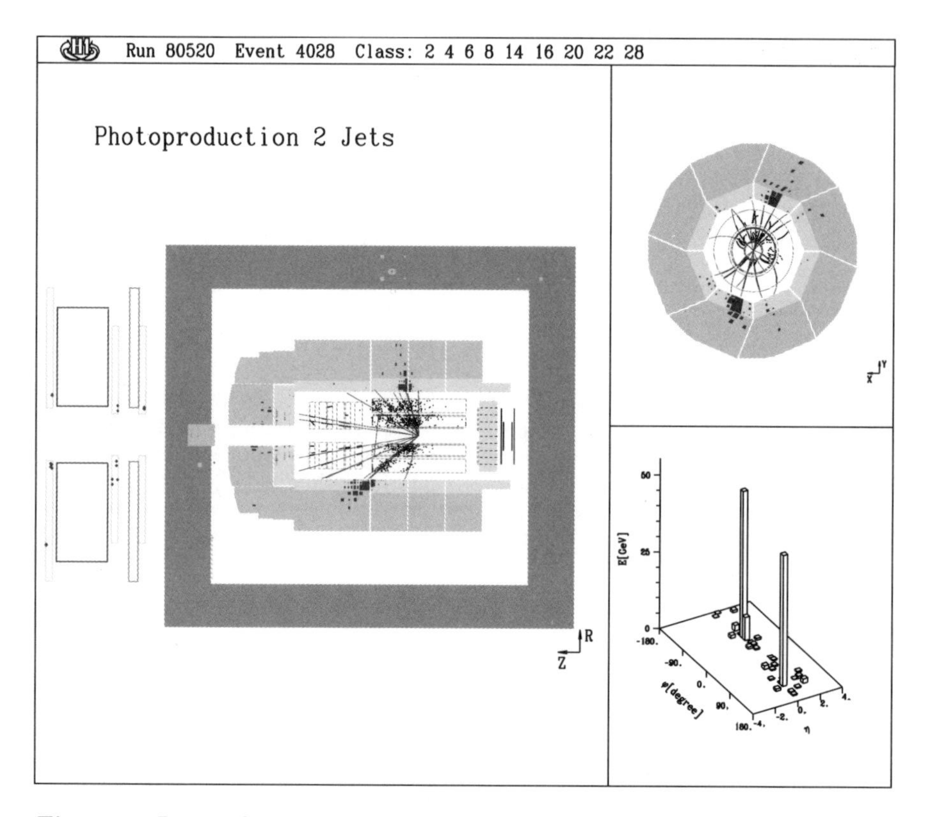

Fig. 4.12. Direct photoproduction of two jets: the side view of the H1 detector is shown on the **left**, where the proton comes from the right and the electron from the left side. Some particles of the proton spectator are shown close to the beam pipe on the left side of the detector. The electron scattered through a small angle and was not detected. The two jets in the central part of the detector are back-to-back in the plane transverse to the beam axis (**upper right**). The histogram shows the energy distributions in the pseudo-rapidity and azimuthal directions (**lower right**)

Before we draw quantitative conclusions from the comparison of jet rates in measured data and QCD calculations, we need to consider the following three topics:

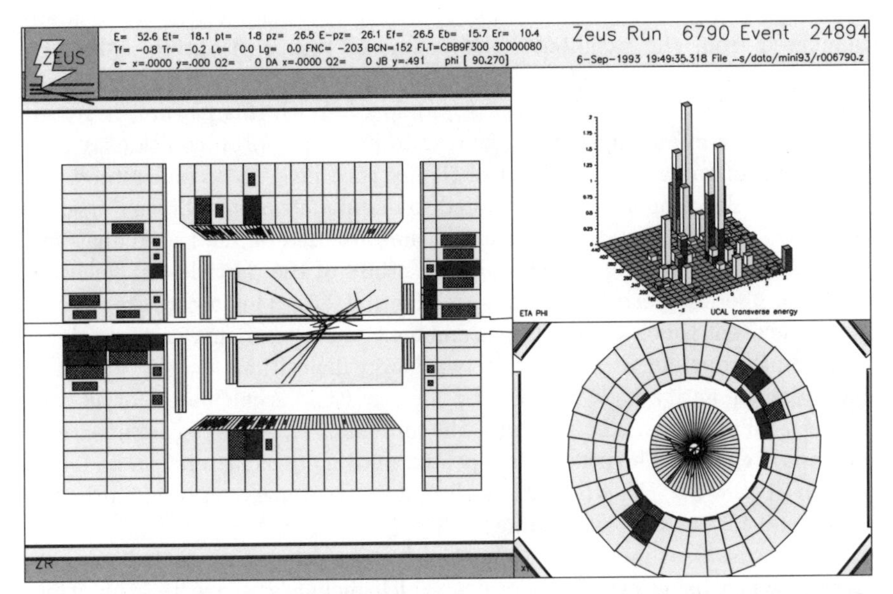

Fig. 4.13. Resolved photoproduction of jets: the figure on the **left** shows the side view of the ZEUS detector, with two jets at large polar angles θ going to the left and a third jet in the scattered electron direction (to the right), which results from the photon remnant. The two jets in the central part of the detector are back-to-back in the plane transverse to the beam axis (**lower right**). The histogram shows the energy distributions in the pseudo-rapidity and azimuthal directions (**upper right**)

1. Jets from γp collisions were reported for the first time at HERA [60, 130]. Are these focused particle bundles of the same jet type as observed in, e.g., $\overline{\text{p}}$p scattering?
2. The jets depend on their definition: experimentalists and theorists need to agree on a jet algorithm.
3. Are the jets described by QCD calculations?

Jet Algorithms. Several algorithms for the jet definition have been studied in γp interactions (e.g., [95]), which can be classified as 'cone' and 'cluster' algorithms.

Most of the HERA photoproduction analyses use a cone algorithm for the jet reconstruction, which follows the recommendations of the 'Snowmass accord' [73]. Here a two-dimensional grid is used in the plane of the pseudo-rapidity η and the azimuthal angle φ, requiring a minimum energy deposit in one of the cells of the grid. The size of the cone is typically chosen to be $R = \sqrt{\Delta\eta^2 + \Delta\varphi^2} = 1$, or $R = 0.7$. The transverse jet energy is simply calculated from the sum of the transverse energies found inside the cone $E_t^{\text{jet}} = \sum_i E_t(\eta_i, \varphi_i)$. The jets are only accepted if the summed energy is above some threshold, which is typically chosen at around

8 GeV at HERA. The rapidity and azimuthal positions of the jet axis are calculated from the weighted energy sums of the contributing cells, e.g., $\eta^{\rm jet} = \sum_i (E_t(\eta_i, \varphi_i) \times \eta_i)/E_t^{\rm jet}$. The cone algorithm cuts out the transverse energy of a clearly defined area in (η, φ) space. With this feature, it is well suited for jets in an environment similar to hadron–hadron collisions, where migrations of energy into and out of the jet cone need to be understood.

Cluster algorithms are better suited to finding jets at small transverse energies and at small angles θ. Such algorithms are used for dedicated analyses, like, e.g., the K_t jet algorithm [30] for a study of the particles fragmenting from the photon spectator partons (see Sect. 4.2.3). This algorithm decomposes the event topology into large combined 'clusters' of energy depositions: based on the opening angle between two energy depositions E_n, E_m and their energy, the quantity $K_t = \min(E_n^2, E_m^2)(1-\cos\theta_{n,m})$ is calculated for all two-cluster combinations. The pair with the minimum K_t value is combined into a common cluster. The process is repeated until no two clusters can be found with K_t below $\xi^{\rm cut} E_t^2$, where E_t is the total transverse energy found in the event, and $\xi^{\rm cut}$ is a cut-off parameter.

Comparisons of measured and calculated jet cross sections are straightforward when using event generators, which include the effects of multiple parton interactions and the hadronization phase. Comparisons of jet data with next-to-leading order QCD calculations are more difficult since the experimental jets are formed from hadrons and the calculated jets are based on partons. The accuracy of the latter comparisons has recently been studied for di-jet cross sections and is better than 20% at transverse jet energies above $E_t^{\rm jet} = 10$ GeV [26].

Properties of Jets in γp, $\bar{\rm p}$p and $\gamma\gamma$ Collisions. We first study the shape of jets from photoproduction events and compare them with theoretical predictions. Then we test a simple scaling law, which relates jets from γp collisions to jets from $\bar{\rm p}$p and $\gamma\gamma$ scattering.

The Shape of Jets from γp Scattering. In Fig. 4.14a the uncorrected transverse energy flow in the azimuthal direction $\Delta\varphi$ around the jet with the highest $E_t^{\rm jet}$ in events with at least two jets is shown from H1 data [68]. The jets were found in the photon fragmentation region between $-2 \le \eta_{\gamma{\rm p}}^{\rm jet} \le -1$, where detector corrections are small. The transverse jet energies are around $E_t^{\rm jet} = 10$ GeV summed in a cone of size $R = 1$. The energy flow was integrated in a slice of $|\eta_{\gamma{\rm p}} - \eta_{\gamma{\rm p}}^{\rm jet}| \le 1$ around the jet axis. The two jets usually deviate from a back-to-back configuration in the azimuthal angle $|\Delta\varphi^{\rm jets}| = \pi$. Here $\Delta\varphi^{\rm jets}$ was defined such that $-\pi \le \Delta\varphi^{\rm jets} \le 0$.

The profile of the jets has approximately Gaussian shape and is described by an empirical fit (full curve), which can be used to determine three features of the jet profiles: the amplitude, the width Γ and the energy level of the underlying event (also called the 'jet pedestal') [68]. The results of a QCD calculation, including fragmentation effects and a detailed simulation of the

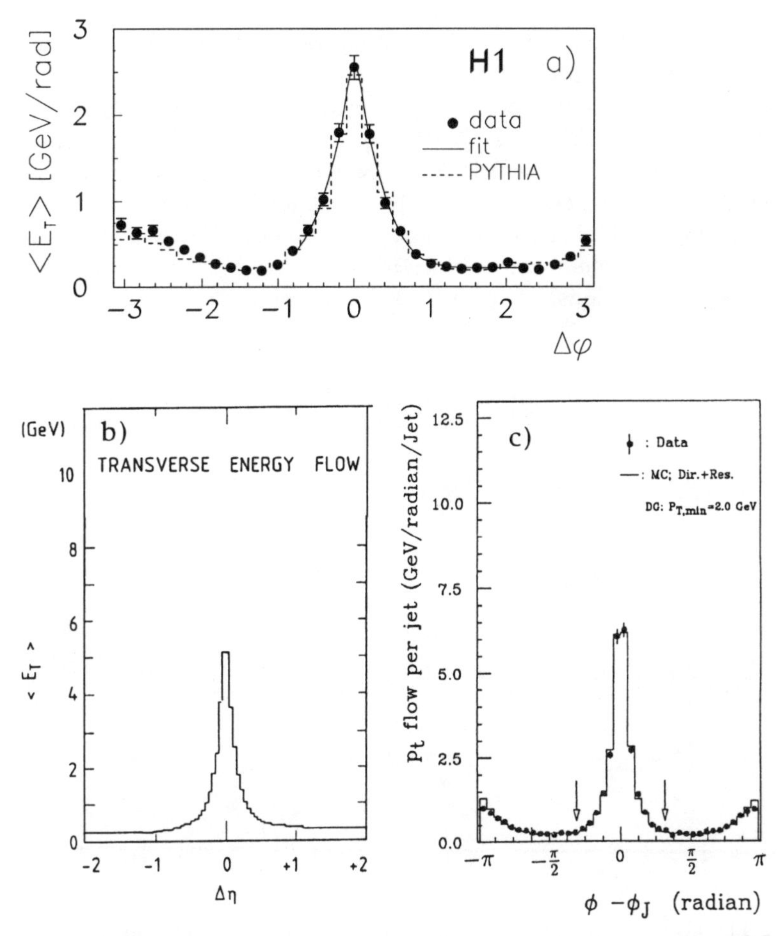

Fig. 4.14. Observed transverse energy flow around a jet axis from different processes: γp, $\bar{\text{p}}$p and $\gamma\gamma$ collisions. (**a**) Photoproduction of jets with transverse energy $E_t^{\text{jet}} = 10$ GeV is shown from H1 data in a projection onto the azimuthal direction (*circles*) [68]. The jets have rapidities between $-2 \leq \eta_{\gamma\text{p}}^{\text{jet}} \leq -1$. The transverse energy of a second jet is seen at $\Delta\varphi \sim \pm\pi$. The axis of this jet had to be at $\Delta\varphi \leq 0$. The latter requirement allows us to measure the energy outside of jets at $\Delta\varphi = 1.5$. The *full curve* is an empirical fit to the data. The *dashed histogram* is a calculation of the QCD generator PYTHIA including a detailed simulation of the detector effects. (**b**) Jet shapes measured in proton–anti-proton collisions are shown in the rapidity projection for jets between $30 \leq E_t^{\text{jet}} \leq 40$ GeV (*histogram*) [118]. (**c**) The azimuthal projection of jet profiles are shown for jets above $E_t^{\text{jet}} \geq 2.5$ GeV, which were observed in the collisions of two quasi-real photons (*circles*) [113]. The *full histogram* is the calculation of a LO QCD generator (reprinted with kind permissions from Springer Verlag and Elsevier Science - NL, Sara Burgerhartstraat 25, 1055 KV Amsterdam, The Netherlands)

detector effects (dashed histogram: PYTHIA), give a satisfactory description of the profile in this rapidity region.

This agreement is no longer present in the central rapidity region of the γp collision $\eta^{\rm jet}_{\gamma p} \approx 0$ (Fig. 4.15 from H1 data [98]). Here the PYTHIA generator without multiple interaction predicts too little transverse energy flux outside the jet area (dotted histogram). This problem cannot be cured by changing the fragmentation of the hard parton scattering process, but is understood as an effect of beam remnant interactions (dashed histogram: PYTHIA with multiple parton interactions), which will be discussed in Sect. 4.3.5.

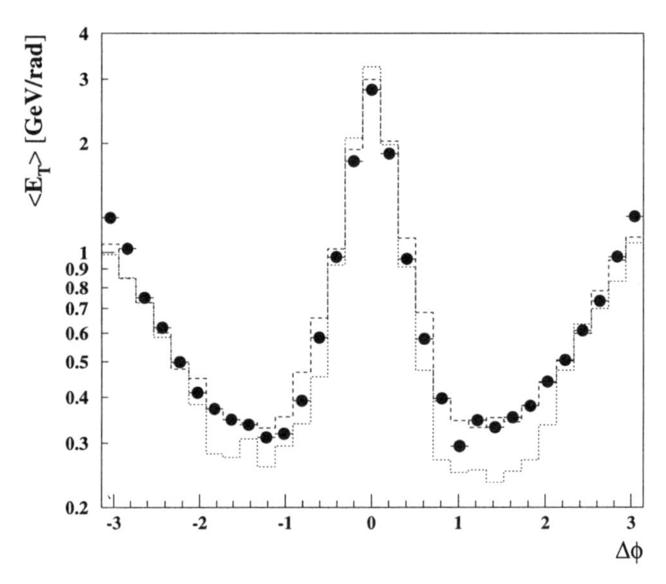

Fig. 4.15. The observed transverse energy flow around the jet axis for jets with transverse energies between $10 \leq E_t^{\rm jet} \leq 12$ GeV is shown in the central rapidity region of the γp collision $0 < \eta^{\rm jet}_{\gamma p} < 0.5$ (*circles*: H1 data [98]). The *dotted histogram* is the calculation of the QCD generator PYTHIA for single hard parton scattering processes per event including a detailed simulation of the detector effects. This calculation cannot describe the measured energy flow next to the jets at $\Delta\varphi = 1.5$. The *dashed histogram* shows the calculation of the same generator including multiple parton scattering processes, which gives an improved description of the measured jet shape

QCD Predictions for Jet Shapes. Two predictions on jet profiles result from QCD calculations on the parton level, which include higher order effects [83]:

1. The width of jets should decrease with increasing jet energy. In the $\Delta\varphi$ projection, the decrease can be approximated by a $1/E_t^{\rm jet}$ behavior.
2. The jets of γp and $\overline{\rm p}$p with equal scaled transverse jet energy

$$x_T = \frac{2E_t^{\rm jet}}{\sqrt{s}} \qquad (4.1)$$

should have the same jet shape. The authors recommend that the normalization on the center-of-mass energy should be $\sqrt{s_{\bar{p}p}}$ for the proton–anti-proton data, and $\sqrt{s_{ep}}$ for the photoproduction data.

Both the simple scaling behavior (4.1) and the dependence of the jet width on E_t^{jet} are phenomenological interpretations of the calculated results.

Comparisons of the Jet Shapes from γp, $\bar{p}$p and $\gamma\gamma$ Processes. In Fig. 4.16 the fitted full width Γ at half maximum above the pedestal is shown for jets between $-2 \leq \eta_{\gamma p}^{jet} \leq -1$ as a function of $x_T = 2E_t^{jet}/\sqrt{s_{ep}}$ from H1 data (full points [68]). The measured jets become narrower with increasing jet energies, and can be fitted by the functional form $1/E_t^{jet}$ (curve).

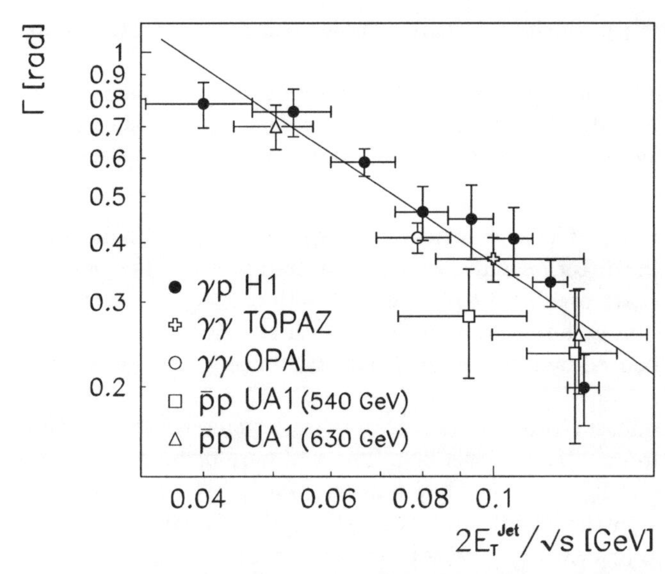

Fig. 4.16. The jet width Γ was determined at half maximum above the energy level measured next to the jets. The width is shown as a function of the ratio of the transverse jet energy over the collision energy. The *full circles* are the results of fits to H1 photoproduction data [68] where the jet energies were normalized to the ep center-of-mass energy, following [83]. The *full line* is a fit to the data proportional to $1/E_t^{jet}$, which is the functional form expected from QCD calculations. The *open squares* and *triangles* show the results of the same fits applied to jet shapes measured in $\bar{p}$p collisions [118, 120]. The *open cross* shows a fit to the TOPAZ $\gamma\gamma$ data (Fig. 4.14 [113]) and the *open circle* a fit to the OPAL $\gamma\gamma$ data [106]

Published results on jet profiles from $\bar{p}$p experiments [118, 120] (e.g., Fig. 4.14b) have been fitted with the same fitting function [68]. The resulting jet widths are shown in the same Fig. 4.16 as open squares and triangles for different jet energies and beam energies. Within the error bars, the widths of the γp jets are compatible with the jet widths determined in the $\bar{p}$p data at the same value of x_T.

In Fig. 4.14c, a profile of jets from the collisions of two quasi-real photons is shown for transverse energies above $E_t^{\rm jet} \geq 2.5$ GeV [113]. This jet profile was also fitted with the same fitting function. The width of these jets is shown in Fig. 4.16 by the open cross at $x_T = 2E_t^{\rm jet}/\sqrt{s_{\rm ee}}$. As the average transverse jet energy, we use here $E_t^{\rm jet} = 3$ GeV. Also shown is a fit to the jet profiles found in $\gamma\gamma$ collisions with the OPAL experiment (open circle [106]). The jets from $\gamma\gamma$ collisions show a shape which is compatible with the simple scaling law (4.1).

Summary

1. The jets observed in photoproduction events show an approximately Gaussian shape.
2. Narrowing of the jet width with increasing jet transverse energies is observed.
3. The jets resulting from γp, $\bar{\rm p}$p and $\gamma\gamma$ scattering have similar shapes. Their widths are related by a simple scaling law depending on the center-of-mass beam energy. These results support the idea that jets are universal.

Transverse Jet Energy Cross Section. Jets with sufficiently high transverse energy reflect the direction and energy of scattered partons. The parton–parton cross sections are predicted to fall as p_t^{-n} with $n = 4$. The transverse jet energy cross section is, however, convoluted with the parton distribution functions and is influenced by fragmentation effects, which jointly modify the power n.

In Fig. 4.17 the measured differential transverse jet energy cross section $\mathrm{d}\sigma/\mathrm{d}E_t^{\rm jet}$ is shown from a H1 analysis [68]. The jets were defined by a cone algorithm with cone radius $R = 1$. They have transverse jet energies above $E_t = 7$ GeV and are in the rapidity interval $-1 < \eta^{\rm jet} < 1$. The event kinematics correspond to photon virtualities below $Q^2 = 0.01$ GeV^2, and a large interval in the photon–proton center-of-mass energy $150 < \sqrt{s_{\gamma \rm p}} < 250$ GeV. The measurement is based on 1993 data with an integrated luminosity of 0.29 pb^{-1}. The data can be described by a power law distribution $(E_t^{\rm jet})^{-n}$ with $n = 6.3$, where the power n is larger than $n = 4$, owing to the convolution with the input distributions.

In the same Fig. 4.17, the data are compared to a QCD calculation of the PHOJET generator (full curve). Here the GRV-LO parton distribution functions were used for the proton and the photon [49, 51]. The calculation gives a good description of the data.

Summary

As in the case of particle production, the measured transverse jet energy cross section is well described by the QCD matrix elements together with standard parton distributions for the photon and the proton.

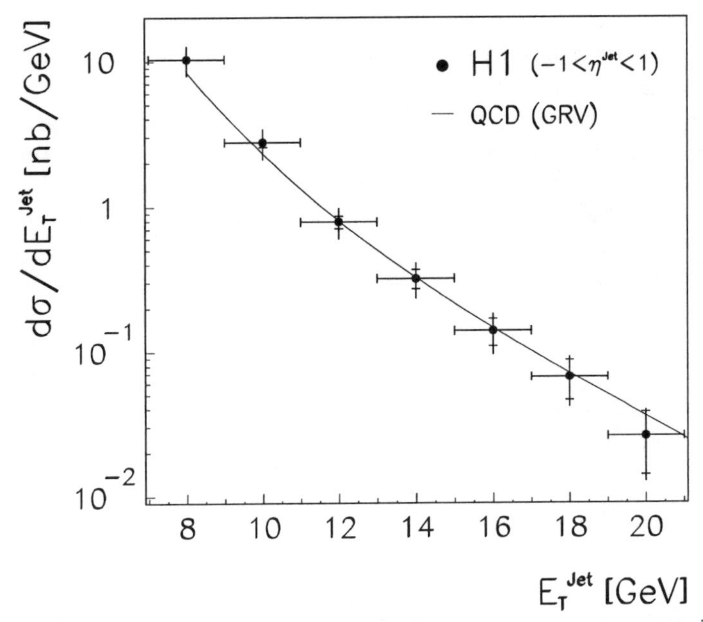

Fig. 4.17. The inclusive differential jet ep cross section $\mathrm{d}\sigma/\mathrm{d}E_t^{\mathrm{jet}}$ is shown as a function of the transverse jet energy from H1 data (*symbols*)[68]. The jet rapidity range was between $-1 \leq \eta^{\mathrm{jet}} \leq 1$. The event kinematics corresponds to $150 \leq \sqrt{s_{\gamma\mathrm{p}}} \leq 250$ GeV and $Q^2 < 0.01\ \mathrm{GeV}^2$. The *full curve* represents a QCD calculation of the PHOJET event generator using the GRV-LO parton distribution functions for the proton and the photon [49, 51]

4.2.3 Resolved Photon Interactions

Resolved γp interactions can be identified by the measurement of the particles from the photon remnant. Here we want to check the existence of the photon remnant and study its properties. Since the remnant particles result from partons, they should form a jet. Such jets would then reflect the intrinsic transverse momenta k_t^γ of the partons from the photon. The k_t^γ distribution is expected to be broader than that of hadrons because of the anomalous component of the photon.

Transverse energy flow. The photon remnant energy should be large for small parton fractional energy x_γ, and should vanish in the case of direct γp interactions ($x_\gamma \equiv 1$). Therefore the study of the x_γ dependence of the transverse energy flow in the photon direction should reveal the photon remnant.

In leading-order QCD, x_γ can be calculated from the two scattered partons using (2.41). A good estimate of x_γ is given by the corresponding jet quantities in events with two jets:

$$x_\gamma^{\mathrm{jets}} = \frac{E_t^{\mathrm{jet1}}\mathrm{e}^{-\eta^{\mathrm{jet1}}} + E_t^{\mathrm{jet2}}\mathrm{e}^{-\eta^{\mathrm{jet2}}}}{2E_\gamma} \tag{4.2}$$

In Fig. 4.18 the transverse energy flow is shown as a function of the rapidity difference $\Delta\eta$ relative to jets with rapidity between $0 < \eta^{\rm jet} < 1$ (from H1 di-jet data [74]). The jets had transverse energies above $E_t^{\rm jet} > 7$ GeV. The photon energy E_γ was determined from the energy of the scattered electron, which was measured in the luminosity detector (2.38). In order to exclude the energy of the second jet, energy deposits were only integrated in the azimuthal range $|\varphi - \varphi^{\rm jet}| < 1$. The data are shown in two intervals of $x_\gamma^{\rm jets} > 0.45$ and $x_\gamma^{\rm jets} < 0.4$.

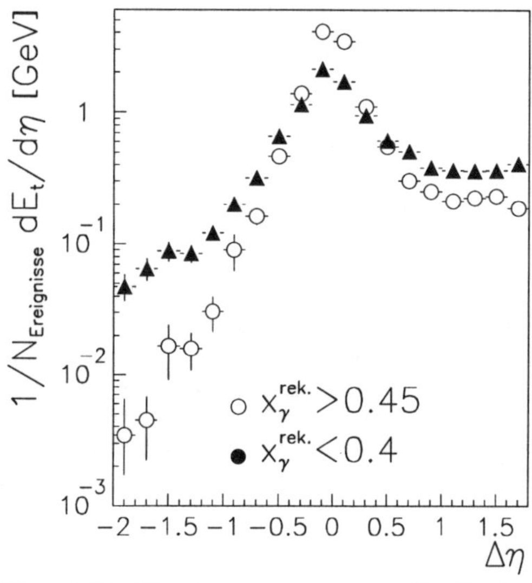

Fig. 4.18. The transverse energy distribution next to the jet with the highest transverse jet energy in the event is shown as a function of the rapidity distance from the jet axis in two intervals of the parton fractional energy $x_\gamma^{\rm jets}$: $x_\gamma^{\rm jets} > 0.45$ (*circles*), $x_\gamma^{\rm jets} < 0.4$ (*triangles*) (H1 data, [74]). The jet was at rapidity $0 < \eta^{\rm jet} < 1$. A second jet was required with transverse jet energy above $E_t^{\rm jet} = 7$ GeV. The transverse energy was integrated in the azimuthal slice $|\varphi - \varphi^{\rm jet}| < 1$ in order to exclude the energy of a second jet in the event, which is needed to calculate $x_\gamma^{\rm jets}$ (4.2)

At low $x_\gamma^{\rm jets}$ the transverse energy flow in the photon direction ($\Delta\eta < 0$) is larger by one order of magnitude compared with the high $x_\gamma^{\rm jets}$ data. This enhanced energy flow shows the existence of the photon remnant, and, therefore, the presence of resolved photon interactions.

In the proton direction ($\Delta\eta > 0$), the transverse energy flow is also enlarged, here by a factor two, for the low $x_\gamma^{\rm jets}$ data relative to the high $x_\gamma^{\rm jets}$ events. This observation is explained by initial-state parton radiation and multiple parton interaction effects (Figs. 4.24, 4.29).

The Photon Remnant Jet . Particles resulting from the photon spectator partons should form a jet. The ZEUS detector is well suited to study hadronic energy depositions close to the beam pipe in the γ direction. The following analysis is based on three jet events from an integrated luminosity of 0.55 pb^{-1} in the kinematic range $Q^2 < 4\ \mathrm{GeV}^2$ and $130 \leq \sqrt{s_{\gamma \mathrm{p}}} \leq 270$ GeV [136].

In order to detect a potentially small transverse energy deposition, the K_t jet algorithm was used to decompose the event into three so-called energy 'clusters' (see Sect. 4.2.2). If two clusters had transverse energies above $E_t^{\mathrm{cluster}} \geq 6$ GeV and were at rapidities below $\eta^{\mathrm{jet}} \leq 1.6$, the third cluster was found to be on average in the photon fragmentation region, well away from the central γp collision region. Such configurations can be interpreted as the result of a hard resolved γp scattering process: the two leading E_t clusters reflect the jets from the hard parton scattering, and the third cluster is the energy deposition resulting from the photon remnant.

The energy distributions along and around the axis of the third cluster were found to be well collimated. In Fig. 4.19a the longitudinal and transverse energy distributions with respect to the cluster axis are shown for this cluster as a function of the cluster energy E_3 (full points). A QCD calculation of the PYTHIA generator for single hard parton interactions gives a good description of this energy distribution (open triangles). In Fig. 4.19b the same quantities are shown for all three clusters as a function of the cluster energy (points for the photon remnant, stars for the high E_t clusters). Within the error bars the third cluster shows the properties of a jet, as one would expect from the partonic picture of photon–proton interactions.

Transverse Momenta of the Partons from the Photon. Owing to the anomalous photon contribution, the photon can split into a $\mathrm{q\bar{q}}$ pair, which can have large transverse momenta k_t^γ relative to the γ direction. The distributions of k_t^γ are expected to follow a power law, rather than a Gaussian distribution, which one would expect for partons from hadrons [34, 40, 102, 103]. The k_t^γ distribution should be reflected in the photon remnant jet distributions.

In Fig. 4.20a the rapidity distribution of the third cluster is shown between $-4 \leq \eta_3 \leq 2$ from the ZEUS analysis. The region around $\eta_3 \sim 2$ corresponds to the central γp collision region. The distribution was corrected for detector effects, including energy losses in the beam pipe, using an unfolding procedure [36]. The full histogram in the figure represents the calculation of the PYTHIA QCD generator, using here a Gaussian, i.e., hadronic, distribution of the parton momenta relative to the photon direction. This calculation does not describe the data. The dotted histogram shows the same QCD calculation, using a power law type k_t^γ distribution of the partons in the photon: $\mathrm{d}N/\mathrm{d}k_t^2 \propto ((k_t^\gamma)^2 + k_\circ^2)^{-1}$ with $k_\circ = 0.66$. This calculation gives a fair description of the data.

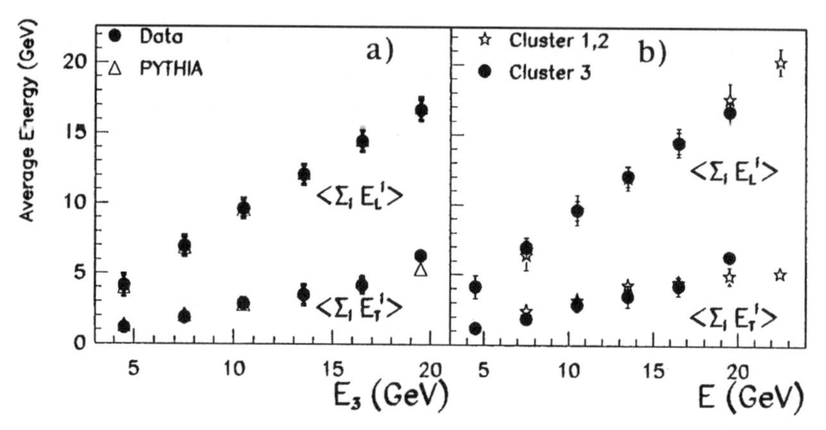

Fig. 4.19. (**a**) The longitudinal and transverse energy distributions with respect to the axis of the photon remnant cluster are shown from ZEUS data (*full circles*). The *open triangles* represent the calculations of the PYTHIA generator. (**b**) The same photon remnant energy distributions are compared with the two jets from the hard scattering process (*stars*), which demonstrates that the photon remnant forms a jet (reprinted from [136] with kind permission from Elsevier Science NL, Sara Burgerhartstraat 25, 1055 KV Amsterdam, The Netherlands)

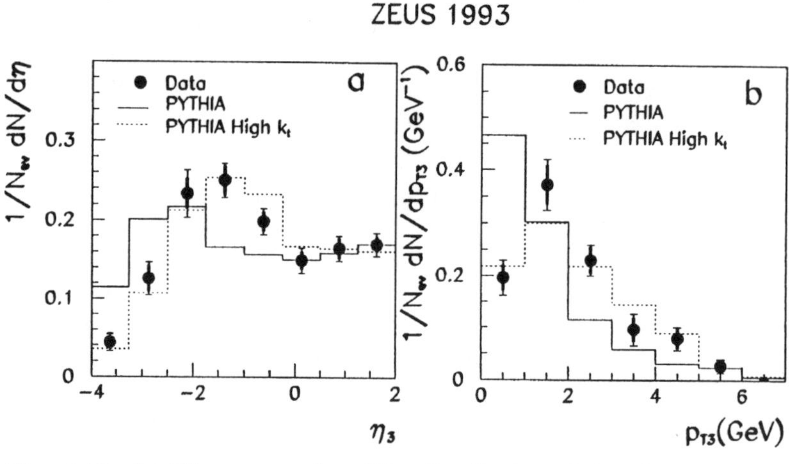

Fig. 4.20. (**a**) The corrected rapidity distribution of the photon remnant jet is shown from ZEUS data (*symbols*). The *full histogram* represents the QCD calculation of the PYTHIA generator using a hadronic type distribution (Gaussian) for the parton transverse momenta k_t^γ from the photon. The *dotted histogram* shows the same calculation with a power-law-like distribution for k_t^γ. (**b**) The corrected transverse momentum distribution of the photon remnant jet is shown for jet rapidities below $\eta_3 < -1$. Histogram assignment is as in (a) (reprinted from [136] with kind permission from Elsevier Science - NL, Sara Burgerhartstraat 25, 1055 KV Amsterdam, The Netherlands)

In Fig. 4.20b the transverse momentum distribution of the third cluster is shown for rapidities below $\eta_3 \le -1$. While in the central γp collision region $\eta \approx 2$ such jets may originate from QCD radiation effects, these clusters are safely inside the photon fragmentation region. Compared with a calculation using the Gaussian k_t^γ distribution, the measured clusters have larger transverse momenta. Also here the calculation with the power-law-like k_t^γ distribution is compatible with the data.

Summary

1. The measurement of photon remnant particles in jet events demonstrates the presence of resolved γp interactions.
2. Studies of the energy flow distribution in the photon fragmentation region show that the photon spectator partons form a jet.
3. The transverse energy $E_t^{\rm jet}$ of the photon remnant jet reflects the transverse momentum k_t^γ of the partons from the photon. The measured $E_t^{\rm jet}$ distribution is much harder than expected from a hadronic type k_t^γ distribution and reflects the anomalous photon component.

4.2.4 Direct Photon Interactions

The lifetime of the photon fluctuations into $q\overline{q}$ pairs is finite according to the uncertainty relation (2.3, 2.8). So, the photon must also exist in its bare state when interacting with a nucleon.

Two methods have been used for the search for direct photon–nucleon interactions:

1. QCD theory predicts a relative suppression of the direct photon processes with respect to the resolved photon processes at small parton scattering angles $\hat{\theta}$ in the parton–parton center-of-mass system, i.e., small transverse parton momenta $\hat{p}_t$ (Fig. 2.6). Therefore, searches for deviations from resolved photon processes are most promising at large transverse parton momenta.
2. In the case of direct photon interactions, the full photon energy enters the parton scattering process. Studies of the distribution of the parton fractional energy x_γ should therefore show an enhancement at $x_\gamma = 1$.

Results of Fixed Target γN Experiments. The fixed target experiments of the 1980s applied a nice trick: with the same detector they measured photon–nucleon and meson–nucleon scattering and compared the production of pions at large transverse momenta p_t. The NA14 experiment found a clear excess of π° production in γN scattering compared with meson–N interactions [91].

A similar excess of γp data was found in the production of charged hadrons by the WA69 experiment. This measurement is shown in Fig. 4.21a [92]. The difference in the measured cross sections could be well explained by QCD calculations of direct photon–nucleon interactions (Fig. 4.21b). According to these results the photon can be decomposed into a vector meson and a direct component. No sign of the anomalous photon component, which was established by deep inelastic lepton–photon scattering experiments (Fig. 3.3), was found at the center-of-mass energy $\sqrt{s_{\gamma p}} = 18$ GeV.

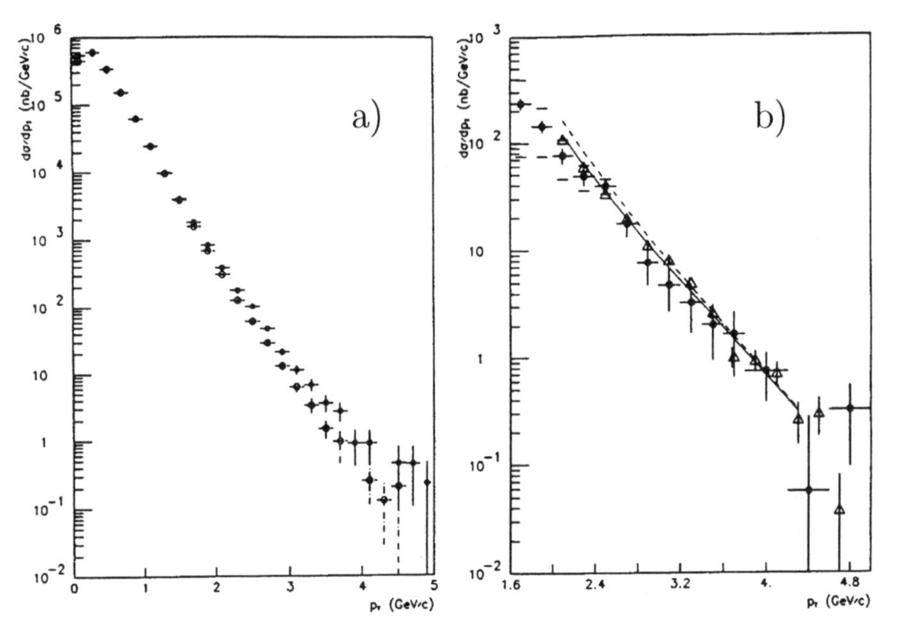

Fig. 4.21. (**a**) The differential cross sections of charged particle production in fixed target photon–proton (*full circles*) and meson–proton (*open circles*) scattering are compared as a function of the transverse particle momenta p_t. The meson-proton data were normalized to the photon–proton data at small $p_t < 1$ GeV. (**b**) The difference between the photon and meson data (*full circles*) is compared with a LO QCD calculation (*open triangles*) and with two NLO QCD calculations: the *full curve* represents the direct photon case, the *dashed curve* includes also the anomalous photon coupling (reprinted from [92] with kind permission from Springer Verlag)

Di-jet Measurements at HERA. At HERA the most interesting results come from the two-jet distributions as a function of the parton fractional energy x_γ. The ZEUS analysis shown here was based on an integrated luminosity of 0.55 pb^{-1} [133]. The scattered electron was required to be absent in the main calorimeter corresponding to a virtuality of the photon below $Q^2 = 4$ GeV^2. The γp center-of-mass energy was between $130 \leq \sqrt{s_{\gamma p}} \leq 270$ GeV. Both jets had transverse energies above $E_t^{\rm jet} \geq 5$ GeV, and rapidities in the interval $-1.125 \leq \eta^{\rm jet} \leq 1.875$. The rapidity difference between the two jets had to be below $|\Delta\eta^{\rm jet}| \leq 0.5$, which selected events with $\approx 90°$ scattering angle in the parton–parton center-of-mass system.

The energy fraction of the parton from the photon x_γ was reconstructed using the two jets with the highest transverse energies in the event:

$$x_\gamma^{\rm cal} = \frac{E_t^{\rm jet1}{\rm e}^{-\eta^{\rm jet1}} + E_t^{\rm jet2}{\rm e}^{-\eta^{\rm jet2}}}{\sum_{\rm hadrons} E_t e^{-\eta}} \tag{4.3}$$

The photon energy was reconstructed using $E_\gamma = (\sum_{\rm hadrons} E_t e^{-\eta})/2$. This gives on average a value that is smaller than the true photon energy, because in resolved photon–proton interactions particles of the photon spectator are lost in the beam pipe. Experimentally, the advantage of (4.3) is that uncertainties in the energy measurement, e.g., event by event fluctuations resulting from energy resolution effects, or uncertainties in the absolute energy scale of the calorimeter, cancel out.

In Fig. 4.22 the two-jet spectrum is shown as a function of the reconstructed energy fraction $x_\gamma^{\rm cal}$ of the parton from the photon. The distribution has two peaks, the large peak at $x_\gamma^{\rm cal} = 0.2$ corresponds to resolved photon–proton-scattering processes, the smaller peak at $x_\gamma^{\rm cal} = 0.8$ is associated with the direct photon–proton interactions. The histograms in the same figure show QCD predictions of the generators HERWIG (dashed) and PYTHIA (full). Both were normalized to the peak at $x_\gamma^{\rm cal} = 0.8$. Also shown is the direct photon contribution to the total rate as calculated by the HERWIG generator, which demonstrates that the peak at large $x_\gamma^{\rm cal}$ is caused by direct photon events.

At small $x_\gamma^{\rm cal} = 0.2$ the energy flow problems of generator programs with only single hard parton interactions per event become apparent. This is not relevant at large $x_\gamma^{\rm cal} > 0.4$ (Fig. 4.29). In the large $x_\gamma^{\rm cal}$ region (Fig. 4.22), the observed data are not purely direct photon interactions, but are 'contaminated' by resolved photon interactions. Therefore, from this distribution alone, it is not possible to make an exact determination of the direct photon cross section, which could improve our knowledge of the parton distributions in the proton. Instead, information on the parton distributions of the photon in the kinematic region $x_\gamma \approx 1$ can be obtained, where the error bars of the deep inelastic lepton–photon scattering experiments are sizable (Fig. 3.1).

Summary

1. The existence of direct photon–nucleon interactions at small photon virtuality, which was demonstrated in fixed target γN experiments, was confirmed in γp collisions at HERA.
2. The QCD prediction for the direct photon cross section is correct at center-of-mass energies $\sqrt{s_{\gamma\rm N}} = 18$ GeV, and is compatible with the HERA data at $\sqrt{s_{\gamma\rm p}} = 200$ GeV, taking into account the uncertainties of the resolved photon contributions.

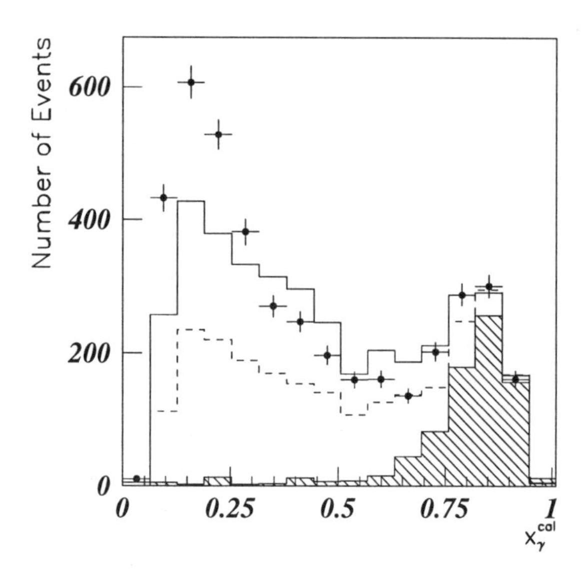

Fig. 4.22. The observed 2-jet spectrum is shown as a function of the fractional energy of the parton from the photon side, using (4.3) (ZEUS Collab.). The histograms represent QCD calculations using the HERWIG (*dashed*) and PYTHIA (*full*) generators for single hard parton scattering processes. The *hatched histogram* shows the direct photon contribution as calculated with the HERWIG generator (reprinted from [133] with kind permission from Elsevier Science - NL, Sara Burgerhartstraat 25, 1055 KV Amsterdam, The Netherlands)

4.2.5 The Parton Scattering Angle

A basic QCD prediction is that different sub-processes have different distributions of the parton scattering angle $\hat{\theta}$. For example, the parton cross section as a function of $\cos\hat{\theta}$ varies much faster for resolved photon interactions than for direct photon processes (Fig. 2.6).

In di-jet events $\cos\hat{\theta}$ can be calculated from the rapidity difference between the two jets (2.37). Direct and resolved processes can be distinguished by the parton fractional energy x_γ (Fig. 4.22).

In Fig. 4.23a, the shape of the measured $\cos\hat{\theta}$ distribution is shown from 1994 ZEUS data [140]. The analysis is based on an integrated luminosity of 2.65 pb^{-1} of the 1994 data-taking period. The event kinematics corresponds to photon virtualities below $Q^2 = 4\ \mathrm{GeV}^2$ and center-of-mass energies between $150 < \sqrt{s_{\gamma\mathrm{p}}} < 270$ GeV. The jets had transverse energies above $E_t^{\mathrm{jet}} = 6$ GeV collected in a cone of size $R = 1$. The rapidity of both jets had to be below $\eta^{\mathrm{jet}} = 2.5$, and the rapidity sum $|\eta^{\mathrm{jet1}} + \eta^{\mathrm{jet2}}| < 0.5$. The invariant di-jet mass was required to be above 23 GeV. The latter cut corresponds to the requirement of large parton–parton center-of-mass energies.

The full symbols in Fig. 4.23a refer to the resolved γp interactions, $x_\gamma < 0.75$, the open symbols reflect the direct γp processes, $x_\gamma > 0.75$. Here x_γ

was calculated using (4.3). Both distributions were normalized to one at $\cos\hat{\theta} = 0$. The measurements show a much steeper rise in the resolved γp angular distribution than in the direct γp processes.

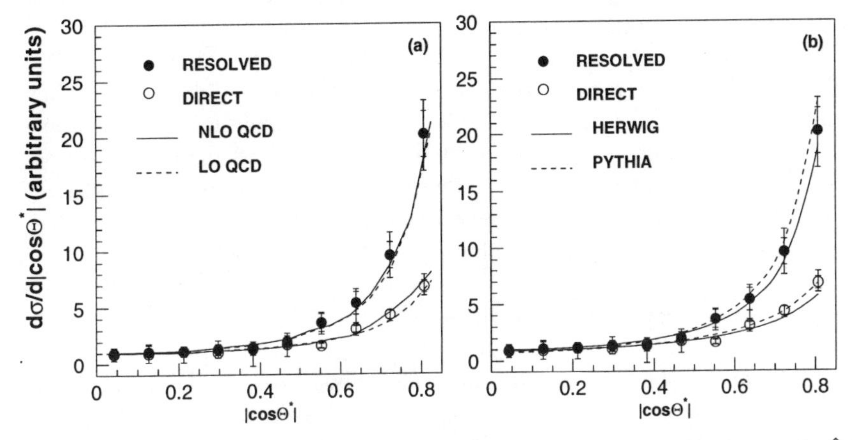

Fig. 4.23. (**a**) The distribution of the cosine of the parton scattering angle $\hat{\theta}$ is shown from di-jet events in two intervals of the parton fractional energy $x_\gamma > 0.75$ (*open symbols*) and $x_\gamma < 0.75$ (*full symbols*) (ZEUS Collab.). Both distributions were normalized to one at $\cos\hat{\theta} = 0$. The curves represent the predictions of analytical LO (*dashed*) and a partial NLO (*full*) QCD calculations [93]. (**b**) The data are compared with the predictions of two LO QCD generators: PYTHIA (*full curve*) and HERWIG (*dashed*) (reprinted from [140] with kind permission from Elsevier Science - NL, Sara Burgerhartstraat 25, 1055 KV Amsterdam, The Netherlands)

In the same Fig. 4.23a, the data are compared with analytical LO (dashed) and partial NLO (full) QCD calculations [93]. Both calculations are compatible with the data. In Fig. 4.23b, the data are compared with the predictions of two generators: PYTHIA (full curve) and HERWIG (dashed). The description of the PYTHIA generator is good; the HERWIG prediction is compatible with the data.

Summary

The basic QCD prediction, that different sub-processes have different distributions of the parton scattering angle, is confirmed in di-jet data.

4.2.6 Higher-Order QCD Effects

Are higher-order QCD corrections to the LO calculation needed to describe the data? Such effects can be studied by the imbalance between the transverse energies of the leading E_t^{jet} jets or by the observation of multi-jet events ($n^{\text{jet}} > 2$).

In Fig. 4.24a the shape of the transverse energy imbalance $\Delta E_t = |E_t^{\text{jet1}} - E_t^{\text{jet2}}|$ between two jets is shown from H1 data [97]. The data show a tail towards large values of ΔE_t. Both jets had transverse energies above $E_t^{\text{jet}} = 7$ GeV, collected in a cone of size $R = 0.7$, and rapidities between $-0.5 < \eta^{\text{jet}} < 2.5$. The reconstructed parton fractional energy x_γ^{jets} (4.2) had to be $x_\gamma^{\text{jets}} > 0.4$ in order to reduce effects of multiple parton interactions (Fig. 4.29). The rapidity distance between the two jets had to be $|\Delta\eta| < 1$ in order to have the same jet pedestal on average. The missing total transverse energy of the events had to be below $E_t^{\text{miss}} = 5$ GeV to ensure that the transverse jet energies were well measured.

The shape of the ΔE_t distribution is described by a PYTHIA calculation with multiple parton interactions, which includes hard initial-state parton showers (full histogram in Fig. 4.24a). The dashed histogram represents the prediction of the PHOJET generator in a version that does not include hard initial-state parton radiation effects: this calculation gives too small ΔE_t. Hard initial-state parton radiation effects are essential to describe the data.

In Fig. 4.24b the transverse momentum imbalance of the two scattered partons is shown for the PYTHIA calculation (full histogram). Their Δp_t distribution results primarily from initial-state parton radiation effects: for comparison, the intrinsic transverse momentum k_t distributions of the partons from the proton (dotted histogram) and the photon (dash-dotted histogram) are also shown.

Fig. 4.25 shows a four-jet event observed with the H1 detector. This is another demonstration that higher-order effects are present.

Summary

1. Hard scattering processes in photon–proton collisions were unambiguously observed by particles with large transverse momenta and jets with high transverse energies.
2. Jets from γp interactions are of the same nature as the jets found in $\overline{\text{p}}$p and $\gamma\gamma$ scattering.
3. The differential jet and particle transverse momentum cross sections at rapidities around $\eta = 0$ are described by QCD calculations.
4. Direct and resolved photon–proton processes were both observed in events with at least two jets.
5. The distributions of the parton scattering angle were found to be different for direct and resolved γ interactions. They are correctly predicted by QCD.

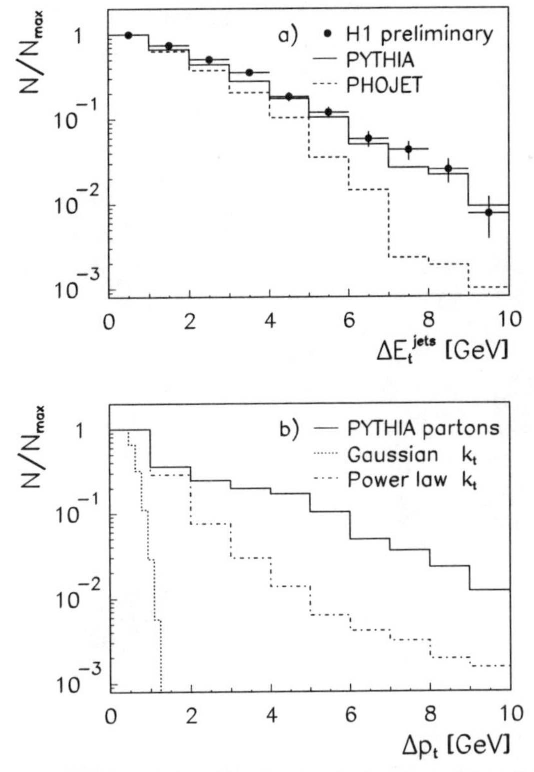

Fig. 4.24. (**a**) The shape of the uncorrected jet transverse energy imbalance $\Delta E_t = |E_t^{\text{jet1}} - E_t^{\text{jet2}}|$ is shown from H1 di-jet data (*full symbols* [97]). The jets had transverse energies above $E_t^{\text{jet}} = 7$ GeV. The reconstructed parton fractional energy had to be $x_\gamma^{\text{jets}} > 0.4$ in order to reduce effects of multiple parton interactions. The *full* and *dashed* histograms show the predictions of the PYTHIA and PHOJET generators including a detailed simulation of the detector effects. The PYTHIA calculation provided hard initial-state parton radiation effects; the PHOJET calculation did not. (**b**) The *full* histogram reflects the transverse momentum imbalance of the scattered partons of the PYTHIA calculation. This imbalance results primarily from parton radiation effects: the *dotted* histogram indicates the shape of a Gaussian transverse momentum k_t^{p} distribution of the partons from the proton. The *dash-dotted* histogram reflects the shape of a power law k_t^γ distribution of the partons from the photon

6. The presence of higher-order processes was demonstrated in di-jet events by the imbalance of the transverse jet energies and by the observation of multi-jet events.
7. Jet formation from the photon spectator partons was observed. The transverse jet energy distribution reflects the transverse momenta of the partons from the photon, which were found to be large, as expected from the anomalous photon component.

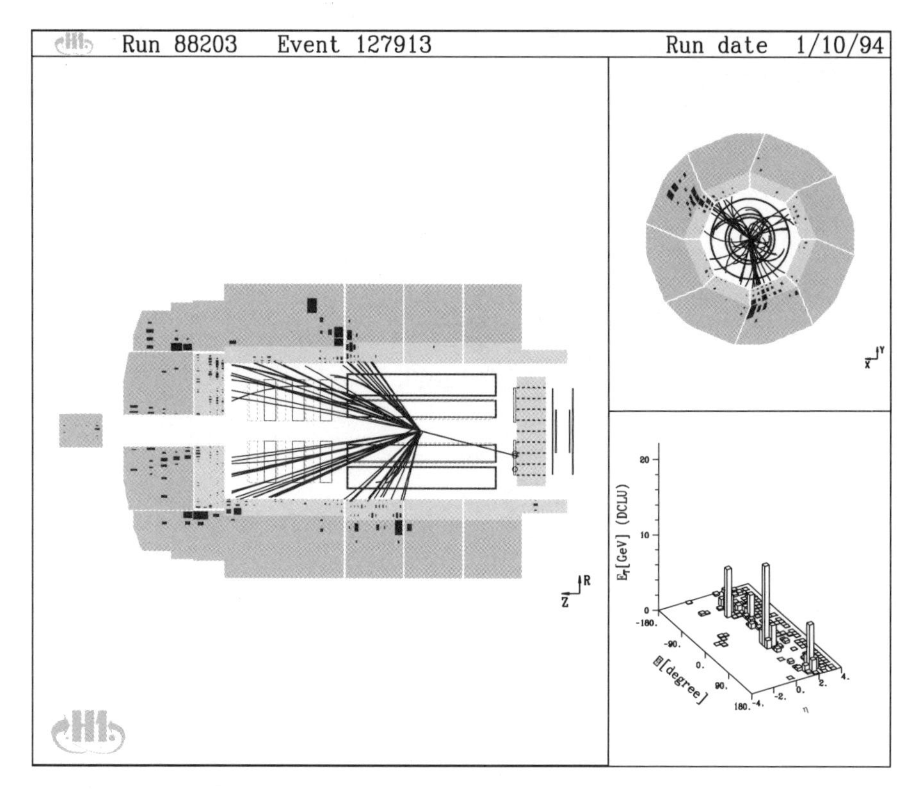

Fig. 4.25. Photoproduction of four jets: the side view of the H1 detector is shown on the **left** side, where the proton comes from the right, the electron from the left. Some particles of the proton spectator are shown close to the beam pipe on the left side of the detector. The electron was scattered through a small angle and was detected in the electron detector of the luminosity system. The **upper right** picture shows the plane transverse to the beam axis. The histogram shows the transverse energy-weighted jet positions in the pseudo-rapidity and azimuthal directions (**lower right**)

4.3 Multiple Parton Interactions

Jets with large transverse energy reflect the properties of the scattered partons. The parton rapidity and azimuthal direction can be precisely determined from the measured jets. The determination of the parton transverse energy through the jet measurement is more difficult. In the picture of the cone jet algorithm, the energy differences are understood as migrations into and out of the cone:

1. Energy of the parton from the hard scattering process falls *outside* of the cone:
   a) *Higher-order QCD effects*: the scattered parton radiates other partons (final-state parton shower).

   b) *Parton fragmentation*: the hadrons originating from the partons have a non-zero angle with respect to the parton direction.
2. Apart from a hard scattered parton, the following effects contribute energy *inside* the cone:
   a) *Higher-order QCD effects*: the initial-state partons radiate other partons before the hard scattering process.
   b) *Multiple parton interactions*: interactions between the two beam remnants spread energy, which appears uncorrelated with the primary hard scattering process.
   c) *Remnant fragmentation*: hadrons resulting from the non-interacting beam remnants appear at large angles with respect to the beam direction.

Except for item 2b), all the effects mentioned above are provided in standard QCD generators. The existence and importance of remnant interactions has yet to be demonstrated and has developed into a separate field of research. It has been approached in two different aspects: at pp and $\overline{\text{p}}$p colliders, the extreme case of simultaneous double parton scattering has been searched for. We summarize their results in the following section.

At HERA, multiple parton interactions are turned on or off depending on the state of the photon. Only resolved γp processes can have interactions between the beam remnants, since no photon remnant exists in direct photon interactions. In the sections on the HERA results, the effects of multiple parton scattering are quantified studying transverse energy distributions with respect to the following questions:

1. How much transverse energy is produced in the events?
2. How is the transverse energy distributed along the γp collision axis?
3. How strongly correlated are the energy depositions?
4. How much transverse energy flow appears next to jets?

4.3.1 Double Parton Scattering in pp and $\overline{\text{p}}$p Collisions

Here we will briefly discuss the results of pp and $\overline{\text{p}}$p experiments on double parton scattering, which leads to the production of four jets.

The cross section σ_{dp} for such double parton scattering can be calculated from the di-jet cross section $\sigma_{\text{di-jet}}$ times the probability of having have a second di-jet production in the same event:

$$\sigma_{\text{dp}} = \sigma_{\text{di-jet}} \frac{\sigma_{\text{di-jet}}}{2\,\sigma_{\text{eff}}} . \tag{4.4}$$

σ_{eff} is the part of the total cross section that is relevant to di-jet production. The factor 2 has been suggested in [109]. For uncorrelated double-di-jet production, σ_{eff} is proportional to the total nondiffractive cross section

$$\sigma_{\text{eff}} = \frac{\sigma_{\text{nondiff.}}}{2} . \tag{4.5}$$

The rate of double parton scattering is highly suppressed with respect to the double bremsstrahlung effects, which give four parton final states from higher order QCD processes. Therefore, observables were designed that are sensitive to the differences between double parton interactions and double bremsstrahlung effects. Such observables study the jet configurations in the azimuthal direction. An example is shown in Fig. 4.26a from [32]: in the case of double parton scattering, the transverse momenta of each di-jet pair are well balanced, and the difference in the azimuthal production angle Δ_S between the two di-jet pairs is uncorrelated, i.e., the distribution of Δ_S is flat. In contrast to this, the double bremsstrahlung processes give a Δ_S distribution that peaks at $\Delta_S = \pi$.

The AFS collaboration reported evidence for double parton interactions in four-jet events with $E_t^{\text{jet}} \geq 4$ GeV at the center-of-mass energy $\sqrt{s_{\text{pp}}} = 63$ GeV [3]. This result was not confirmed by a study of the UA2 collaboration, which used four jets with $E_t^{\text{jet}} > 15$ GeV in $\bar{\text{p}}$p collisions at $\sqrt{s_{\bar{\text{p}}\text{p}}} = 630$ GeV [122]. They set a limit on the normalizing cross section $\sigma_{\text{eff}} > 8.3$ mb at 95% confidence level.

More recently, evidence for double parton scattering was reported by the CDF collaboration using jet energies above $E_t^{\text{jet}} = 25$ GeV at the center-of-mass energy $\sqrt{s_{\bar{\text{p}}\text{p}}} = 1800$ GeV [32]. In Fig. 4.26b the measured distribution of the two di-jet pairs is shown as a function of the azimuthal angle Δ_S between the di-jet pairs. The histogram represents the result of QCD double bremsstrahlung processes. The difference between the dotted curve and the histogram gives the contribution of double parton scattering processes, which was found to be at the level of 7% with respect to the double bremsstrahlung processes. The double parton cross section was determined to be $\sigma_{\text{dp}} = 63\pm^{32}_{28}$ nb for parton transverse momenta $p_t > 18$ GeV corresponding to an effective cross section $\sigma_{\text{eff}} = 12.1\pm^{10.7}_{5.4}$ mb. This result is compatible with uncorrelated double parton scattering (4.5), because the nondiffractive cross section amounts to $\sigma_{\text{nondiff.}} \approx 44$ mb. It is also compatible with the limit set by the UA2 experiment.

These measurements indicate that several parton–parton collisions can happen in one event. However, the four-jet analyses are complicated, because one searches for a small effect within large hadronic activity. Therefore, the results need support from other measurements. If multiple parton interactions are universal, they should also be found in processes other than $\bar{\text{p}}$p interactions, e.g., in resolved γp collisions at HERA.

4.3.2 Inclusive Transverse Energy Cross Section

On the HERA measurements, we first study the transverse energy of the final-state hadrons, independently of jet production. The H1 analysis, described here, was based on a 1993 sample of quasi-real photon–proton collisions ($Q^2 \leq 0.01$ GeV^2) at average center-of-mass energies $\sqrt{s_{\gamma\text{p}}} = 200$ GeV [68].

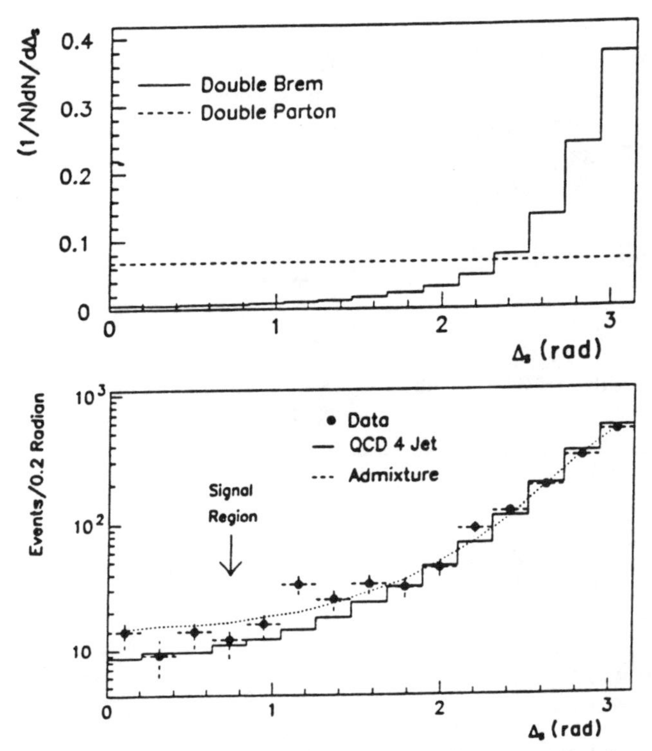

Fig. 4.26. Four-jet production in $\overline{\rm p}$p collisions results from QCD double bremsstrahlung and double parton scattering processes. (**a**) Two jet pairs were formed such that each pair gave an optimal back-to-back configuration. The observable $\Delta_{\rm S}$ then describes the difference in the azimuthal angle between the di-jet pairs. Since double parton scattering results from two essentially uncorrelated scattering processes, the distribution in $\Delta_{\rm S}$ should be flat. In contrast, the distribution of di-jet pairs from double bremsstrahlung are expected to peak at $\Delta_{\rm S} = \pi$ (from [32]).
(**b**) The *symbols* show the CDF measurement of four-jet events as a function of $\Delta_{\rm S}$. The histogram gives the predicted distribution of the double bremsstrahlung processes. The dotted curve represents the sum of the double bremsstrahlung processes and the contribution of double parton scattering processes

The data correspond to an integrated luminosity of 0.29 $\rm pb^{-1}$. In Fig. 4.27a the differential transverse energy cross section is shown for transverse energies between $20 \le E_t \le 50$ GeV. The energy was summed in the rapidity region $-2.5 \le \eta_{\gamma \rm p} \le 1$ and was corrected for detector effects. The errors are dominated by the uncertainty in the knowledge of the absolute H1 calorimeter energy scale. The cross section is around 10 nb/GeV at $E_t = 20$ GeV and shows an exponentially decreasing distribution $\exp(-\lambda E_t)$ with $\lambda = 0.21 \pm 0.01$. Such exponential transverse energy distributions have been observed in pp and $\overline{\rm p}$p collisions at sufficiently high energy before [2, 31, 119].

The data in Fig. 4.27a could equally well be fitted by a power law E_t^{-n}, which one would naively expect from parton scattering processes. A fit to

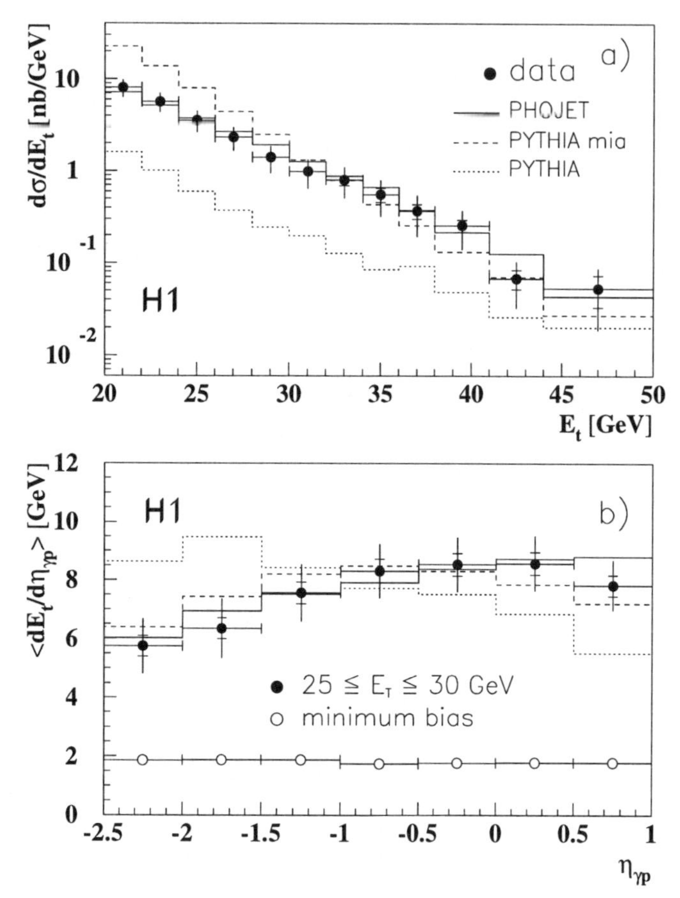

Fig. 4.27. (**a**) The inclusive differential transverse energy cross section $\mathrm{d}\sigma/\mathrm{d}E_t$ is shown for the γp center-of-mass rapidity interval $-2.5 \leq \eta_{\gamma\mathrm{p}} \leq 1$ from H1 data (*symbols*). The *dotted* and *dashed histograms* show the QCD calculation of the PYTHIA generator for single hard parton scattering processes and for multiple parton interactions respectively. The *full histogram* is a calculation of the PHOJET generator, which also includes multiple parton interactions. (**b**) The transverse energy is shown as a function of $\eta_{\gamma\mathrm{p}}$ for the total transverse energies between $25 \leq E_t \leq 30$ GeV (*full points*). The measurement is compared with the energy level found in nondiffractive soft scattering processes (minimum bias data: *open circles*). Histogram assignment is as in (a) (reprinted from [68] with kind permission from Springer Verlag)

the data gave $n = 5.9 \pm 0.1$, which is close to the power law found in the inclusive jet cross sections (Fig. 4.17) and supports the interpretation of this distribution in terms of hard parton scattering processes.

Also shown in the same figure are calculations of different QCD generators, using the same parameterizations of the parton distributions in the proton and the photon throughout (GRV-LO [49, 51]). The PYTHIA single parton interaction model is far below the data, and shows a power law behavior, rather than an exponential distribution (dotted histogram). The PYTHIA calculation with multiple parton interactions shows instead an exponentially decreasing distribution, with a different slope compared with the data (dashed histogram). In this model relatively small energy depositions of spectator interactions are added on top of the energy from the hard parton scattering process. They not only give a higher cross section, since more events get over the E_t threshold, but also cause a significant change in the shape of the distribution. The PHOJET predictions (full histogram), which include multiple parton scattering, were found to be well compatible with the data in the measured E_t range. The main difference between the PYTHIA and PHOJET models is that PYTHIA includes only QCD hard multiple parton interactions, while PHOJET contains soft and hard multiple parton scattering processes.

4.3.3 Transverse Energy Flow

Figure 4.27b shows the average transverse energy per event as a function of the γp center-of-mass rapidity. For this measurement, the total transverse energy was restricted to the interval $25 \leq E_t \leq 30$ GeV (Fig. 4.27a). The data were corrected for detector effects, where the corrections depend slightly on the generator model used and give the dominant measurement error in this energy distribution. The energy flow per event in minimum bias data (open circles) is also shown in the figure. This data set is defined by a trigger condition which accepts 95% of the total nondiffractive photoproduction cross section (Fig. 2.9). While the minimum bias data are perfectly flat in this rapidity interval, as one would expect from soft hadronic interactions, the high-energy data differ from such a flat distribution and show a maximum close to the mid-rapidity region (Fig. 4.5). These phenomena are well known from hadron–hadron collisions: a) flat rapidity distributions have been extensively studied at the ISR [22, 29], while b) at the large beam energies of $\bar{\text{p}}$p collider experiments the rapidity distributions with sufficiently high transverse energy peak at mid-rapidity [119].

The observation that the transverse energy distribution of the γp data is not flat means that these data cannot result from soft γp interaction, but show the scattering of constituents that have less than the full beam energy. Naively, one would expect this distribution to be peaked somewhere in the photon fragmentation region, i.e., at negative rapidities, since more partons with large fractional energy x exist in the photon than in the proton. In the same Fig. 4.27b the PYTHIA calculation for one hard parton interaction per event is shown, which supports this simple view (dotted histogram). Possible explanations for the maximum at mid-rapidity are:

(a) The distribution is dominated by the scattering of partons with small fractional energy x and the parton distributions of the photon and proton are similar. This would require a large gluon content in the photon.

(b) The effect could be explained without requiring a large gluon content in the photon by the superposition of several parton scattering processes with a fraction of processes at small x to fill energy into the γp center-of-mass region.

The latter scenario is implemented in PYTHIA with hard multiple parton interactions, which is shown in the figure as the dashed histogram. The maximum is shifted relative to the calculation of the single parton interaction model towards the central rapidity region of the γp collision, but the calculation still does not coincide with the data. The PHOJET calculation gives a fair description of these data distributions.

Summary

The description of the measured transverse energy distributions is much improved by the inclusion of multiple parton interactions.

4.3.4 Energy–Energy Correlations

In previous figures (e.g., Figs. 4.15, 4.27), QCD generators that allow for multiple parton interactions describe the data significantly better than generators with one parton scattering process per event. In multiple parton interaction models, the parton–parton scattering processes are essentially independent of each other. Here we want to test this model assumption by studying the transverse energy–energy correlations.

Figure 4.28 shows the transverse energy–energy correlations in H1 data (event selection was described in Sect. 4.3.2) in the rapidity projection with respect to the central collision region of the γp center-of-mass system, where most of the energy is concentrated (Fig. 4.27). The H1 detector is well suited for this measurement, since the calorimeter has a fine granulation in the region of interest. The correlation function Ω was defined as

$$\Omega(\eta_{\gamma\mathrm{p}}) = \frac{1}{N_{\mathrm{ev}}} \sum_{i=1}^{N_{\mathrm{ev}}} \frac{\left(\langle E_{t,\eta_{\gamma\mathrm{p}}=0}\rangle - E_{t,\eta_{\gamma\mathrm{p}}=0}\right)_i \left(\langle E_{t,\eta_{\gamma\mathrm{p}}}\rangle - E_{t,\eta_{\gamma\mathrm{p}}}\right)_i}{(E_t^2)_i} . \quad (4.6)$$

The average energies $\langle E_{t,\eta_{\gamma\mathrm{p}}}\rangle$ were determined from all events. The data shown were not corrected for detector effects and have a total transverse energy in the rapidity region shown of $E_t^{\mathrm{vis}} \geq 20$ GeV. Short-range energy correlations are observed close to the mid-rapidity region. According to the definition of the correlation function this means: if the energy deposited at $\eta_{\gamma\mathrm{p}} = 0$ in one event is above the event average energy, the energy in the neighboring bin is also above the event average in that bin. Anti-correlations are observed in the photon hemisphere at rapidity $\eta_{\gamma\mathrm{p}} \sim -1.8$.

The PYTHIA calculation of single hard parton interactions including a detailed simulation of the H1 detector, shows the same shape of the correlation function (dotted histogram). However, the correlation strength between

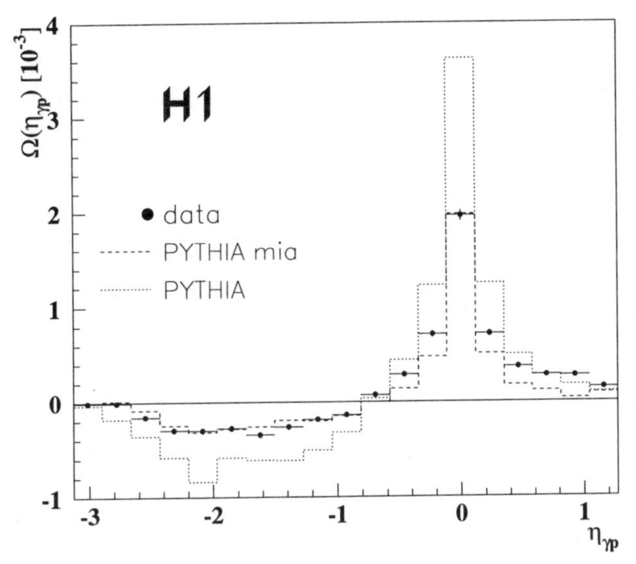

Fig. 4.28. Observed transverse energy–energy correlations with respect to the γp center-of-mass system are shown as a function of the rapidity $\eta_{\gamma p}$ (*circles* from H1 data. The *dotted histogram* shows the QCD calculation of the generator PYTHIA for single hard parton scattering processes, which gives too strong energy–energy correlations. The *dashed histogram* represents the calculation of the same generator including multiple parton interactions, which shows a correlation strength similar to that observed in the data (reprinted from [68]) with kind permission from Springer Verlag)

the transverse energy depositions is twice as large as in the data! In contrast, the same calculation of the PYTHIA generator with multiple interactions (dashed histogram) not only gives a fair description of the shape of the rapidity correlation, but also results in the correct energy correlation strength, as observed in the data. These conclusions do not change with the added requirement of two jets with transverse energy above $E_t^{\text{jet}} > 7$ GeV in the event.

Summary

The studies of energy–energy correlations give strong support to multiple parton interaction models which spread *uncorrelated* energy into an event, in addition to the energy resulting from a hard scattering process.

4.3.5 Underlying Event Energy

If the large measured transverse energy flow results from additional interactions between the spectator partons of the proton and the photon, we expect differences in the energy flow between resolved and direct photoproduction processes: only the resolved photon processes can have spectator interactions, not the direct photon processes, which by definition do not have photon-spectator partons. More generally speaking, we expect a dependence on the energy fraction x_γ of the parton from the photon, which determines the energy E_γ^{sp} remaining for the photon spectator partons: $E_\gamma^{\text{sp}} = (1 - x_\gamma)E_\gamma$. The parton fractional energy from the proton side is typically $x_{\text{p}} \approx 0.01$, so the energy of the proton spectator partons is essentially the same as the proton energy.

Transverse Energy Density at Mid-Rapidity. Since we are looking for effects beyond the primary hard parton scattering process, we need to exclude the transverse energy of the jets themselves. The energy flow, measured outside of jets, is called the *underlying event energy.* From Figs. 4.27b and 4.28, we know that the effects of spectator-parton interactions should be largest at mid-rapidity, where the energy flow is largest and the correlations of the transverse energy depositions are relatively weak.

In Fig. 4.29 the corrected average underlying transverse energy density at mid-rapidity of the γp center-of-mass system is shown for events with at least two jets (points). In addition to the event selection, described in Sect. 4.3.2, two jets were required with transverse energies $E_t^{\text{jet}} > 7$ GeV, collected in a cone of size R=1. The energy was summed in the region $|\eta_{\gamma\text{p}}| \leq 1$. Energy deposited close to the jet axis, within $R \leq 1.3$ of the two jets with the highest transverse energies, was excluded from the energy sum. The summed energy was divided by the area used for the energy summation. Therefore, the vertical axis of Fig. 4.29 represents the energy outside of jets per unit area in (η, φ) space. The horizontal axis represents the parton fractional energy x_γ^{jets} using (4.2). x_γ^{jets} was corrected for detector effects. The error bars of the distribution are dominated by the uncertainty in these corrections.

The energy density of the underlying event in Fig. 4.29 was measured to be 0.4 GeV/rad at large fractional energies x_γ^{jets} and to increase to above 1 GeV/rad at small x_γ^{jets}. The measurement is compared with the energy level found at mid-rapidity in deep inelastic ep scattering events (DIS) at average $\langle Q^2 \rangle = 30$ GeV^2 from H1 data (square), which correspond to 'pure' direct photon–proton interactions [66]. It is interesting to note that the energy level of these DIS events coincides with that of minimum-bias photoproduction data (Fig. 4.27b). Since the only common contribution to the underlying event in these two data sets is the fragmentation process, the initial- and final-state parton radiation in direct photon events cannot have a large influence on the central photon–proton collision region.

The triangle symbols in Fig. 4.29 show the uncorrected underlying event transverse energy measured in $\overline{\text{p}}$p collisions at $\sqrt{s_{\overline{\text{p}}\text{p}}} = 630$ GeV next to jets with transverse energies between $28 \leq E_t^{\text{jet}} \leq 62$ GeV [120]. Such collisions involve initial-state partons with small energies, which we have calculated from $x^{\text{jets}} = E_t^{\text{jet}}/\sqrt{s_{\overline{\text{p}}\text{p}}}$. The energy density is found to be compatible with the energy measured in resolved photon–proton interactions at $x_\gamma^{\text{jets}} = 0.1$

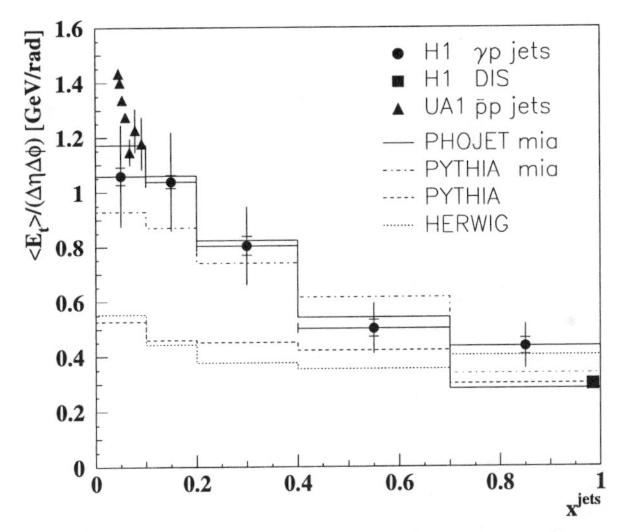

Fig. 4.29. The transverse energy density outside of jets is shown for the central γp collision region $|\eta_{\gamma p}| \leq 1$ from H1 data (*full circles*) [68]. The E_t density was measured as a function of the fractional energy of the parton from the photon side using (4.2). The *histograms* are QCD calculations of the HERWIG (*dotted*) and PYTHIA (*dashed*) generators for one hard parton scattering process per event. The *dash-dotted* and *full histograms* represent the calculations of the PYTHIA and PHOJET generators including multiple parton interactions. The transverse energy density measured in deep inelastic ep scattering (DIS) in the central γ^*p collision region is shown as the *square symbol* for events with average photon virtuality $\langle Q^2 \rangle = 30$ GeV^2 from H1 data [66]. Since this measurement results from direct photon interactions, it has been placed at $x = 1$. The *triangle symbols* represent the raw transverse energy density measured outside of jets in $\overline{\text{p}}$p collisions at $\sqrt{s_{\overline{\text{p}}\text{p}}} = 630$ GeV, where the jets had transverse energies between $28 \leq E_t^{\text{jet}} \leq 62$ GeV (UA1 Collab. [120]). The parton fractional energy was here estimated using $x = E_t^{\text{jet}}/\sqrt{s_{\overline{\text{p}}\text{p}}}$

and tends to increase towards small x^{jets}. This similarity hints at a universal effect.

The amount of initial-state radiation at low x_γ^{jets} has been checked with two QCD generators for single parton scattering processes per event. The results of these calculations are shown in Fig. 4.29 as the dashed (PYTHIA) and dotted (HERWIG) histograms. The calculations show a small increase in the energy density towards small x_γ^{jets}; however, the increase is far below the data. The use of another photon structure function parameterization cannot change this result, since, e.g., a large gluon content at small x_γ^{jets} increases the number of events, but not the average energy density. Attempts to reproduce the large energy density with increased initial-state parton radiation effects did not result in a consistent description of the data.

Fig. 4.29 also shows a calculation of the PYTHIA model with hard multiple parton interactions using the GRV-LO parton distribution functions (dash-dotted histogram). This gives a transverse energy density that is compatible with the data. Also the fluctuations of the energy density are described by this model as well as the jet kinematics at small x_γ [74]. The calculation of this model does depend on the choice of the photon structure function parameterization, because

1. subsequent parton scattering processes result from different parton fractional energy x_γ, and
2. the number of parton scattering processes depends on the gluon density in the photon.

A PYTHIA calculation using the LAC1 parameterization of the photon structure function instead of the GRV-LO set results in an energy density that is too large by 0.5 GeV/rad at small $x_\gamma^{\rm jets}$. A readjustment of the parton transverse momentum cut-off from $\hat{p}_t^{\rm mia} = 1.2$ GeV to $\hat{p}_t^{\rm mia} = 2$ GeV again results in a compatible description of the measured distribution [112]. The calculation of the PHOJET generator, which contains soft and hard multiple parton interactions, gives a fair description of the measured energy density, which is shown as the full histogram in Fig. 4.29 (using the GRV-LO parameterizations).

The Number of Interactions per Event. Independently of the specific multiple interaction model, one can estimate the number of interactions per event between the beam particles via the underlying transverse energy density itself: the calculations of the transverse energy flow $E_t^{\rm sia}$ for single hard parton interactions per event represent the energy resulting from a primary hard scattering process (Fig. 4.29: dashed, dotted histograms). Since distributions of transverse energy fall steeply, we assume that the additional interactions between the beam remnants are not jet-like to first order, but contribute transverse energy $E_t^{\rm min.\ bias}$ at the level of minimum bias events (Fig. 4.27b). With these assumptions, the number of interactions per event can be determined from the measured transverse energy density $E_t^{\rm meas}$:

$$\mathrm{R} = 1 + \frac{\langle E_t^{\rm meas}\rangle - \langle E_t^{\rm sia}\rangle}{\langle E_t^{\rm min.\ bias}\rangle} . \tag{4.7}$$

In Fig. 4.30 the interaction rate is shown as a function of the fractional energy x_γ. We have ignored here the logarithmic dependence of the transverse energy density $E_t^{\rm min.\ bias}$ in minimum bias events on the center-of-mass energy that is available to the spectator system $\sqrt{s} = \sqrt{(1-x_\gamma)s_{\gamma\rm p}}$, because this can only be calculated for the first spectator interaction, not for a subsequent one. Therefore, the rate R is a lower estimate of the interaction rate. The transverse energy density $\langle E_t^{\rm sia}\rangle$ used was the average value of the PYTHIA and HERWIG calculations. The number of interactions per event above $x_\gamma > 0.4$ is compatible with one interaction. Below $x_\gamma < 0.4$, the rate increases to three interactions per event on average between the photon and the proton.

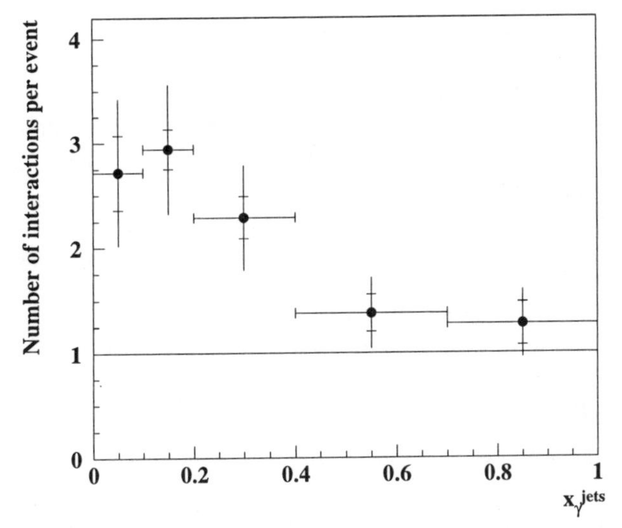

Fig. 4.30. The number of interactions per event was calculated from the transverse energy density of the underlying event: $R = 1 + (\langle E_t^{\text{meas}} \rangle - \langle E_t^{\text{sia}} \rangle)/\langle E_t^{\text{min. bias}} \rangle$. Here $\langle E_t^{\text{meas}} \rangle$ are the measured values from Fig. 4.29. The QCD calculation of the transverse energy density produced by a single hard parton interaction is represented by $\langle E_t^{\text{sia}} \rangle$ (Fig. 4.29: dashed, dotted histograms). The transverse energy density $\langle E_t^{\text{min.bias}} \rangle$, determined in minimum bias photoproduction events, was used as a measure of the transverse energy produced in the interactions between the beam remnants

Summary

1. The measured transverse energy distributions in γp scattering are described much better overall by QCD generators with multiple parton interactions than by generators without them.
2. At large parton fractional energies $x_\gamma > 0.4$, QCD generators, which calculate one hard parton scattering process per event, give a consistent description of the measured transverse energy flow.
3. At low $x_\gamma < 0.4$, such calculations fail to describe the large measured underlying event energy. Analytic next-to-leading QCD calculations also cannot account for this large E_t flow either (P. Aurenche, G.Kramer, private communications 1995). A natural explanation is provided by models that allow for several parton–parton interactions per event. The photon, with its unique feature of direct and resolved photon interactions, gives new insight into multiple parton scattering dynamics, complementary and beyond the double parton scattering reported by pp and $\overline{\text{p}}$p experiments.

4.4 Parton Distributions of the Photon

What is the contribution of particle and jet data in photon–proton and two-photon interactions to our knowledge of the parton distributions in the photon? Do such results go beyond the achievements of the deep inelastic lepton–photon scattering (DIS) experiments?

The results of the DIS $e\gamma$ experiments constrain the quark distribution of the photon to the level of 30% for parton fractional energies between $0.05 < x_\gamma < 0.9$. These data are not sensitive to the gluon distribution of the photon, as was demonstrated by the LAC1–3 parton distribution functions. Uncertainties in the very small, $x_\gamma \ll 0.1$, and very large, $x_\gamma \approx 1$, regions are considerable (see Figs. 3.2, 3.5).

In the following sections we consider five topics:

1. The reason for a lower limit of the parton fractional energy in events with jets or high-transverse momentum particles is explained.
2. Three different procedures are described that are used to extract information on the parton distributions of the photon from the data.
3. Jet and particle measurements are presented, which give information on the quark and gluon distributions of the photon.
4. The influence of the underlying event energy on the extraction of the parton distributions in the photon is analyzed.
5. The contribution of the anomalous photon component to the particle cross section is studied, and the characteristic scale dependence of the quark density in the photon is extracted from di-jet data.

4.4.1 Lower Limit on the Parton Fractional Energy

Particle and jet measurements of photoproduction data at HERA and two-photon data at TRISTAN and LEP are sensitive to the quark and gluon content of the photon (Fig. 2.4). They cover a large region of parton fractional energies up to $x_\gamma = 1$. At small x_γ they are constrained by a lower limit which can be estimated from

$$x_\gamma > \frac{p_t\, e^{-\eta}}{2E_\gamma} \tag{4.8}$$

where

1. Comparisons with perturbative QCD calculations require parton transverse momenta above $p_t \approx 2$ GeV,
2. The rapidity η measurement of a particle or jet is limited by the detector acceptance,
3. The photon energy E_γ is limited by the available beam energies.

Typically, the parton fractional energies are above $x_\gamma = 0.01$ at HERA, and $x_\gamma = 0.1$ at TRISTAN, and $x_\gamma = 0.05/0.01$ at LEP1/2.

Since the parton cross section falls steeply as a function of the parton transverse momentum p_t, predictions on the total rate of hard scattering processes depend on the cut-off value $p_t^{\rm cut}$. Therefore, at the lowest x_γ value mentioned above, conclusions on the parton distributions of the photon depend on $p_t^{\rm cut}$. For conclusions on the absolute density of the partons in the photon, events are selected where the p_t of the hard scattered partons are well above $p_t^{\rm cut}$. With $x_\gamma \propto p_t$ (4.8), this requirement effectively raises the smallest x_γ value reached by the experiments.

4.4.2 Procedures to Extract the Parton Distributions of the Photon

To extract new information on the parton distributions $f_{i/\gamma}$ of the photon from the HERA jet and particle measurements, three different procedures have been applied: a QCD calculation of, for example, jet cross sections requires at least four components and four observables correspondingly (2.28) (see Sects. 2.2.2, 2.2.3). With the set of variables $(y, x_\gamma, x_{\rm p}, p_t)$, the main parts of the calculation are given by

$$\frac{{\rm d}^4\sigma}{{\rm d}y\, {\rm d}x_\gamma\, {\rm d}x_{\rm p}\, {\rm d}p_t} \sim f_{\gamma/{\rm e}}(y) \sum_{ij} f_{i/\gamma}(x_\gamma, p_t^2)\, f_{j/{\rm p}}(x_{\rm p}, p_t^2) \times |M_{ij}\left(\frac{y x_\gamma x_{\rm p}}{p_t^2}\right)|^2 . \tag{4.9}$$

As before, y denotes the scaled photon energy, and x_γ ($x_{\rm p}$) is the fractional energy of the parton from the photon (proton). The factorization and renormalization scales are here chosen to be equal to the jet transverse momentum $\mu^2_{\rm fac} = \mu^2_{\rm ren} = p_t^2$. All components of (4.9), except $f_{i/\gamma}$, are well known: $f_{\gamma/{\rm e}}$ from QED calculations, $f_{j/{\rm p}}$ from measurements of the proton structure function (at typically $x_{\rm p} \sim 10^{-2}$), and the matrix elements M_{ij} from QCD theory.

The three procedures to extract information on the partons $f_{i/\gamma}$ of the photon from the data differ in the treatment of the sum $\sum_{ij}$ over the different initial parton states in (4.9):

1. The cross section measurement is compared directly with the QCD calculation (4.9) using different parameterizations for $f_{i/\gamma}$. The comparison returns the best parameterization of $f_{i/\gamma}$ constructed so far. In future, $f_{i/\gamma}$ will be iterated in order to give the optimal match between the data and the calculation. This procedure is valid to all orders of the QCD calculation.
   Instead of the observable x_γ mentioned in (4.9), the pseudorapidity η of the jets or particles is frequently used together with a cut on their transverse momentum. They are related to the x_γ observable by (2.41). The remaining variables are integrated over.

2. The sum in (4.9) can be separated into the different parton components of the photon, direct photon $f_{\gamma/\gamma}$, quark from the photon $f_{\mathrm{q}/\gamma}$, and gluon from the photon $f_{\mathrm{g}/\gamma}$:

$$\sum_{ij} f_{i/\gamma}\, f_{j/\mathrm{p}}\, |M_{ij}|^2 = f_{\gamma/\gamma} \sum_j f_{j/\mathrm{p}}\, |M_{\gamma j}|^2 + f_{\mathrm{q}/\gamma} \sum_j f_{j/\mathrm{p}}\, |M_{\mathrm{q}\,j}|^2 + f_{\mathrm{g}/\gamma} \sum_j f_{j/\mathrm{p}}\, |M_{\mathrm{g}\,j}|^2 \,. \tag{4.10}$$

   The direct photon processes are predicted by QCD (first term on the right-hand side). For the resolved photon processes, the quark distribution of the photon can be taken from the results of the deep inelastic eγ scattering experiments (second term). The third term contains the gluon density in the photon $f_{\mathrm{g}/\gamma}$. With all other components known, $f_{\mathrm{g}/\gamma}$ can be determined from the data. The procedure is valid in leading order QCD.
3. By defining effective parton distributions in the photon and the proton $\tilde{f} = \sum_{n_f} (\mathrm{q} + \bar{\mathrm{q}}) + (9/4)\,\mathrm{g}$ and by using only a single effective subprocess (SES) for the resolved photon–proton interactions (see Sect. 2.2.2), the sum in (4.9) reduces to the direct γp processes and one resolved γp component:

$$\sum_{ij} f_{i/\gamma}\, f_{j/\mathrm{p}}\, |M_{ij}|^2 \approx f_{\gamma/\gamma} \sum_j f_{j/\mathrm{p}}\, |M_{\gamma j}|^2 + \tilde{f}_\gamma\, \tilde{f}_\mathrm{p}\, |M_\mathrm{SES}|^2 \,. \tag{4.11}$$

   The direct photon interactions are again taken from the QCD prediction. The procedure allows for the extraction of the effective parton distribution $\tilde{f}_\gamma$ in the photon. It is valid in leading order QCD.

4.4.3 Particle Cross Sections

How large are the contributions coming from the different photon components – direct, vector meson (VDM), anomalous – to the measured particle cross sections? Do comparisons of data and QCD calculations give constraints on the quark and gluon distributions of the photon?

In Fig. 4.31 a preliminary H1 analysis [71] of the differential charged particle cross section $\mathrm{d}\sigma/\mathrm{d}p_t$ is shown in comparison with a leading order QCD calculation. The measurement was based on the 1994 data-taking period with an integrated luminosity of 1.3 pb^{-1}. The event kinematics covers photon virtualities $Q^2 < 0.01$ GeV^2, and the scaled photon energy interval $0.3 < y < 0.7$ at average photon–proton center-of-mass energies of $\sqrt{s_{\gamma\mathrm{p}}} = 200$ GeV. The cross section is shown for particles with transverse momenta above $p_t = 2$ GeV, integrated over the rapidity range $|\eta| < 1$.

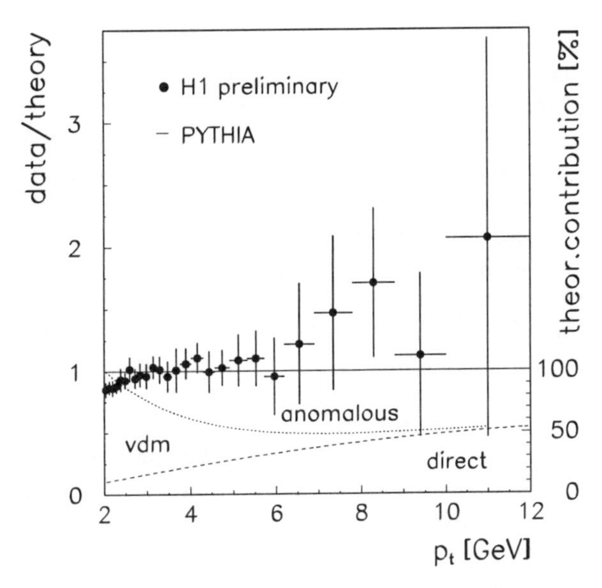

Fig. 4.31. The ratio of the differential particle cross section $(d\sigma^{data}/dp_t)/(d\sigma^{theory}/dp_t)$ is shown from preliminary H1 data (*circles*) [71] and a QCD calculation. The theory part was calculated using PYTHIA with multiple parton interactions and the GRV-LO parton distributions of the proton and the photon [49, 51]. On the right axis, the relative contributions of the different photon components are indicated: the *dashed curve* is the calculation of the direct photon cross section. The *dotted line* indicates the sum of the direct and the vector meson (VDM) photon components, which were normalized to the total calculation at $p_t = 2$ GeV. The VDM contribution was calculated using the GRV-LO parton distribution of the pion instead of the photon. The remaining cross section is assigned to the anomalous photon contribution

The QCD calculation was based on the event generator PYTHIA with multiple parton interactions, using the GRV-LO parameterizations of the parton distributions for the photon and the proton. The data were normalized to the calculation of direct and resolved photon processes. The calculated cross section exceeds the data by 20% at transverse hadron momenta $p_t = 2$ GeV, and is compatible with the measurement above $p_t = 2.5$ GeV.

Also shown in Fig. 4.31 are the relative contributions of the direct, VDM and anomalous photon components. The VDM component was calculated using the parton distribution function of a meson, instead of the photon. Here the GRV-LO parton distribution functions of the pion were used. At small scale $Q_\circ^2 = 0.25$ GeV^2, they have the same quark and gluon distribution as the photon, except for an overall normalization factor. The difference between the parton distributions of the pion and the photon at larger values of the scale results from the anomalous term in the inhomogeneous evolution equation (2.16), which is present only in the case of the photon. The VDM contribution was normalized, together with the direct photon component, to give the total calculated cross section at $p_t = 2$ GeV. The anomalous photon contribution then results from the difference between the total calculated cross section and the VDM plus direct photon components.

In this calculation the anomalous photon contribution amounts to 50% at transverse momenta $p_t > 6$ GeV. Above $p_t = 3.5$ GeV, the photon components, which are predicted by perturbative QCD (direct and anomalous), dominate the VDM contribution. The large center-of-mass energies, available at HERA, give access to the anomalous photon component, which could not be seen in fixed-target experiments so far (Fig. 4.21).

In Fig. 4.32 the measured differential rapidity cross section $\mathrm{d}\sigma/\mathrm{d}\eta$ is shown for particles with transverse momenta above $p_t = 2$ GeV. The corresponding interval of the parton fractional energy can be estimated at between $x_\gamma(\eta = 1) \approx 0.05$ and $x_\gamma(\eta = -1) \approx 0.4$ on average.

The data are compared with predictions of the PYTHIA generator with multiple parton interactions using two different parton distribution functions: GRV-LO (upper histogram) and LAC1 (lower histogram). The prediction using the GRV-LO parameterization gives a cross section that is too large at negative rapidities (large x_γ), and falls short at positive η (small x_γ). The calculation using LAC1 is compatible with the data at $\eta < 0$, but overshoots the measurement at $\eta > 0$.

The dotted histogram indicates the gluon contribution of the photon to the calculation: the steep rise in the gluon distribution, which is tested by the LAC1 parameterization, is not supported by the data.

The particle spectra are insensitive to effects of multiple parton interactions at sufficiently large $p_t > 3$ GeV. Therefore, future comparisons of measured high-p_t-particle cross sections with next-to-leading (NLO) QCD calculations will give an important basis for the extraction of precise NLO quark and gluon distributions for the photon at small and large x_γ [16]. Fragmentation functions of quarks and gluons in NLO, which are needed for the calculations, are obtained from e^+e^- data (Sect. 2.2.2).

Summary

The measured high-p_t-particle cross sections are sensitive to the quark and gluon distributions of the photon.

4.4.4 Single Differential Jet Cross Sections

In this section comparisons of measured single inclusive jet cross sections from $\gamma\gamma$ and γp collisions with QCD calculations will be discussed with respect to the parton distributions of the photon.

Jet Cross Section in $\gamma\gamma$ Scattering. In Fig. 4.33 the differential parton rapidity cross section $\mathrm{d}\sigma/\mathrm{d}\eta$ from the collision of two quasi-real photons is shown from a jet analysis of the AMY experiment [4]. The data correspond

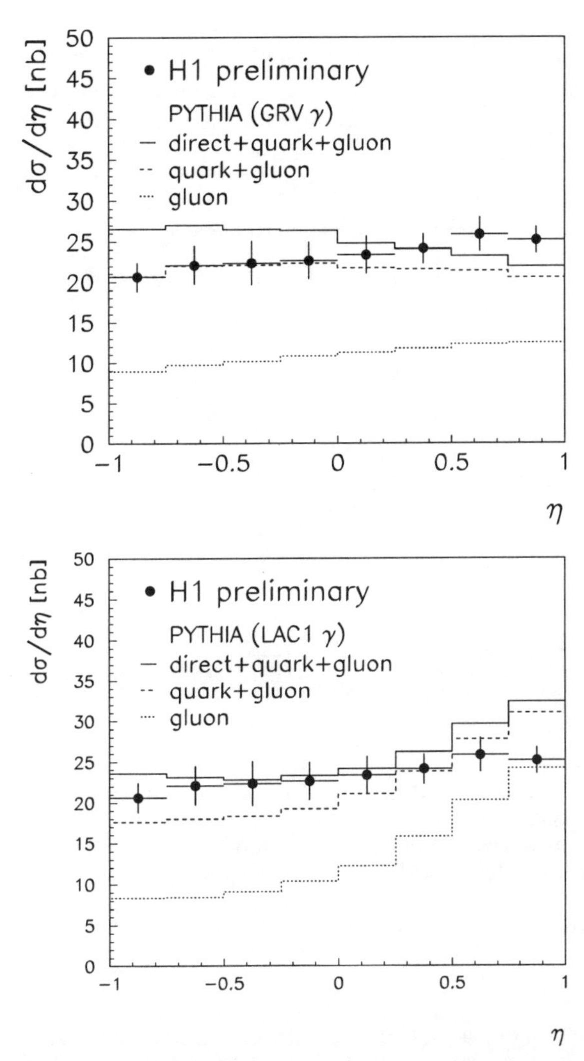

Fig. 4.32. The differential particle cross section $\mathrm{d}\sigma/\mathrm{d}\eta$ is shown as a function of the particle rapidity η for transverse particle momenta above $p_t = 2$ GeV (*symbols*: H1 data [71]). The data are compared with predictions of the PYTHIA event generator with multiple parton interactions using two different photon distribution functions: in the **upper** (**lower**) *full histogram* the GRV-LO (LAC1) parameterizations were used [1, 51]. The *dotted histograms* represent the contribution of the gluon component of the photon. The *dashed histograms* represent the calculations for the resolved photon contributions

to an integrated luminosity of 27.2 pb^{-1}. The virtuality of the photons was below $Q^2 = 10\ \mathrm{GeV}^2$, and the visible two-photon center-of-mass energy between $4 < \sqrt{s_{\gamma\gamma}} < 20$ GeV. The jets had transverse energies in the range $2.5 < E_t^{\mathrm{jet}} < 9$ GeV, collected in a cone of size $R = 1$. The jet data were corrected for detector and fragmentation effects to the level of the scattered partons with transverse momenta $p_t > 2.5$ GeV. The measured cross section is around 80 pb at rapidity $\eta = 0$ and is consistent with being flat over the full measured rapidity range $|\eta| < 1$.

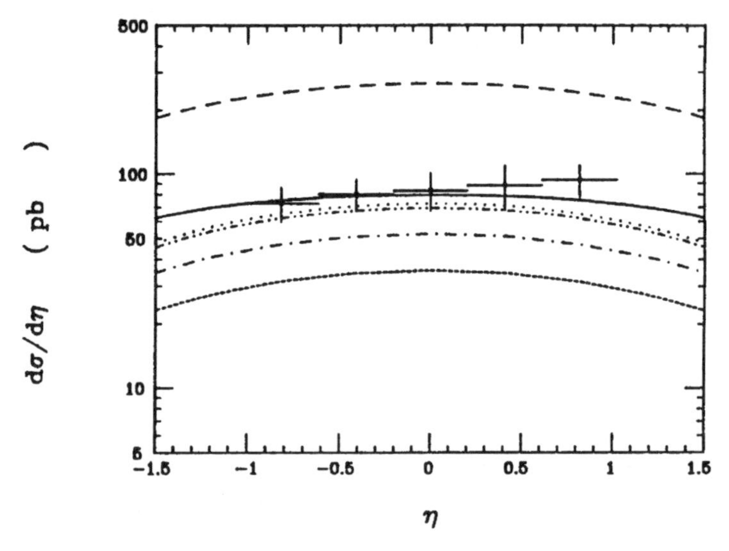

Fig. 4.33. The measured inclusive differential parton rapidity cross section $\mathrm{d}\sigma/\mathrm{d}\eta$ is shown from the collision of two quasi-real photons for partons with transverse momenta $p_t \geq 2.5$ GeV (*symbols*: AMY Collab.). The *short dashed curve* is the calculation of the quark parton model. The other curves are leading order QCD calculations using different parton distribution functions: *long dashed*=LAC3 [1], *full*=LAC1, *dash-dotted*=LAC1 without gluons, *dash-double-dotted*=GRV [51], and *dotted*=DG [39] (reprinted from [4] with kind permission from Elsevier Science - NL, Sara Burgerhartstraat 25, 1055 KV Amsterdam, The Netherlands)

In the same Fig. 4.33, leading order QCD calculations are compared with the data, where different parameterizations of the parton distributions of the photon have been used. The corresponding parton fractional energy x_γ can be estimated from (4.2) to be around $x_\gamma \approx 0.2$ at the scale of $p_t^2 \approx 10$ GeV^2. In this kinematic range, the quark distribution of the photon has been measured by deep inelastic eγ scattering experiments. The direct photon contribution is predicted by QCD (see matrix elements: Table 2.1). These two components together (dash-dotted curve) amount to only 60% of the measured cross section, and show the need for a gluon contribution to the photon. Parameterizations of the parton distributions of the photon, which rely on a gluon component similar to that of mesons, result in QCD calculations which are compatible with the data (full=LAC1 [1], dotted=DG [39], double-dot-dashed=GRV [51]). A large gluon component of the photon at large parton fractional energy x_γ (dashed=LAC3) is excluded. This is compatible with the jet results of γp scattering [61]. The different photon contributions, direct photon (short-dashed), quarks (difference between dash-dotted and short-dashed), and gluons (difference between full and dash-dotted) all have similar shapes. Therefore this measurement is only sensitive to the total jet rate.

Jet Cross Section in γp Scattering at High x_γ. How sensitive are the photoproduction jet data to the quark and gluon distributions of the photon? In this section, we concentrate on the region of large parton fractional energy $x_\gamma > 0.4$ where effects of the underlying event energy are small (Fig. 4.29). This kinematic region essentially corresponds to negative jet rapidities $\eta^{\rm jet} < 0$.

In Fig. 4.34 the differential jet ep cross section $d\sigma/d\eta$ is shown for photoproduction processes from a preliminary ZEUS analysis [129]. Here the rapidity distribution of the 1994 data is shown for transverse jet energies $E_t^{\rm jet} > 17$ GeV. The kinematic range corresponds to $Q^2 < 4$ GeV^2 and $134 \leq \sqrt{s_{\gamma p}} \leq 277$ GeV. The jets were corrected for detector effects, but not for fragmentation or underlying event energy effects. The inner error bars show the statistical uncertainty, and the outer error bars represent the squared sum of the statistical and those systematic errors, which are not associated with the uncertainty in the knowledge of the jet energy scale. The latter error is shown separately in the figure by the shaded band.

The data are compared with an analytical next-to-leading order QCD calculation [77] using the GRV-NLO (full curve) and the GS-96 (dash-dotted curve) parton distribution functions for the photon [49, 54], and the MRSD$^-$ functions for the proton [89]. Contrary to the two-photon calculations, the different photon contributions (dashed=direct γ, dotted=quark, gluon) have very different shapes. The gluon component is here suppressed by the requirement of jets with large transverse jet energy (Fig. 2.7). The region above $\eta^{\rm jet} > 1$ cannot be compared directly, because the effects of the underlying event energy are large in the data and are not included in the QCD calculations. This region will be discussed in the following paragraph. In the region $\eta^{\rm jet} < 0$, the calculation using the GS-96 parton distributions is in agreement with the measurement, while the calculation with the GRV-NLO distributions slightly overshoots the data. Within the experimental errors, both calculations are compatible with the measurement and demonstrate that the quark distributions, which were extracted from the deep inelastic eγ experiments, are universal.

Jet Cross Section in γp Scattering at Small x_γ. Here we want to study the jet cross section at large rapidities $\eta^{\rm jet}$, which correspond to small parton fractional energies x_γ. Two questions are of interest:

1. The QCD calculations of a single hard parton interaction per event fail to describe the underlying event energy (Fig. 4.29). How large is the influence of the underlying event energy on the observed jet rate?

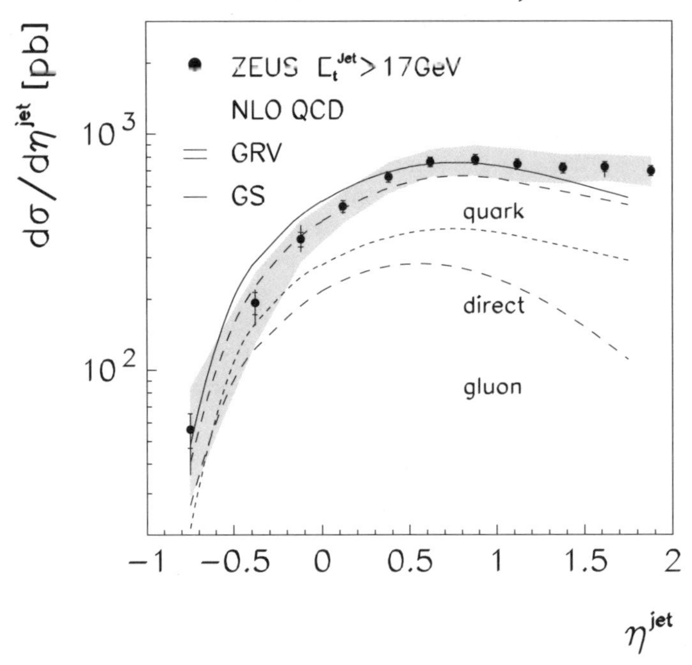

Fig. 4.34. The inclusive differential ep jet-rapidity cross section $d\sigma/d\eta^{\rm jet}$ is shown for transverse jet energies above $E_t^{\rm jet} > 17$ GeV from ZEUS data (*circles*) [129]. The measurements are compared with next-to-leading QCD calculations [77]. The *full (dash-dotted) curve* gives the complete calculation using the GRV-NLO (GS-96) parameterizations of the parton distributions in the photon [49, 54]. The *dashed* curve is the contribution of the direct photon processes. The *dotted curves* show the contributions of resolved photon processes with a quark or a gluon from the photon side

2. Multiple parton interaction models describe the underlying event energy at small x_γ. How sensitive are comparisons between jet data and calculations with respect to the parton distributions of the photon?

Here we first study the calculations of different QCD event generators. Figure 4.35 shows the ratio of two predictions for the differential jet cross section as a function of the jet rapidity. The jets have transverse energies above $E_t^{\rm jet} = 11$ GeV, summed in a cone of size $R = 1$. The event kinematics was determined by an electron tag, where the virtuality of the photon was below $Q^2 < 0.01$ GeV^2 and the photon–proton center-of-mass energy was between $150 < \sqrt{s_{\gamma \rm p}} < 250$ GeV. As the reference, the PYTHIA calculation for a single hard parton interaction per event is used together with the GRV-LO parton distributions in the proton and the photon [49, 51].

The effect of the underlying event energy on the jet rate is studied using two generators, which include multiple parton interactions. The upper edge of the hedged band shows the result of the PYTHIA calculation, and the lower edge represents the calculation of the PHOJET generator. The jet rate is increased by a factor 1.7 at large rapidities $1.5 \leq \eta^{\rm jet} \leq 2$ owing to the underlying event energy. Note that these calculations describe the measured underlying event energy (Fig. 4.29: full, dash-dotted histograms).

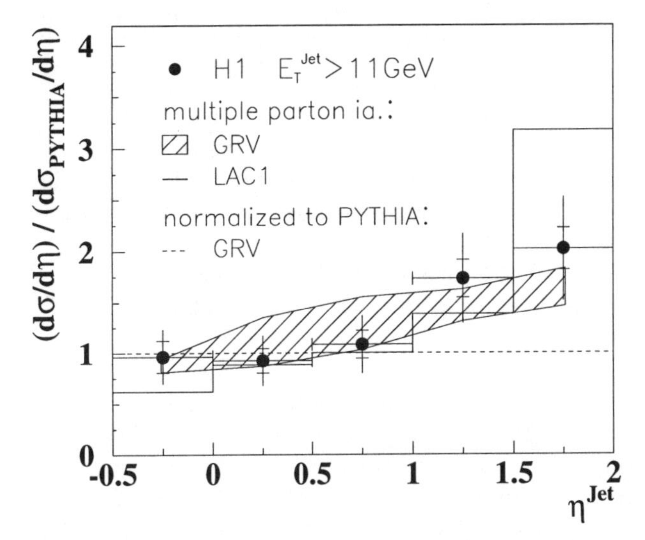

Fig. 4.35. The ratio of the measured jet cross section and a calculation of the PYTHIA generator is shown as a function of the jet rapidity (*full circles*: H1 data [68]). The jets had transverse energies above $E_t^{\rm jet} = 11$ GeV. The PYTHIA version without multiple parton interactions was used together with the GRV-LO parton distribution functions. The *hatched band* shows the predictions of the PHOJET (*lower edge*) and PYTHIA (*upper edge*) generators, both including multiple parton interactions and both using the GRV-LO parton distribution functions. The *histogram* is the prediction of PYTHIA with multiple parton interactions using the LAC1 parton distribution functions

To study the dependence of the jet cross section on the rate of the hard scattering processes, the PYTHIA calculation with multiple parton interactions was repeated using another parton distribution for the photon (Fig. 4.35: histogram). Here the LAC1 parameterization [1] was used, which gives a higher gluon density at small x_γ than that given by the GRV-LO parameterization. The parton transverse momentum cut-off in PYTHIA was raised from $\hat{p}_t^{\rm mia} = 1.2$ GeV (GRV-LO) to $\hat{p}_t^{\rm mia} = 2$ GeV (LAC1) in order to describe the measured underlying event energy (Fig. 4.29). The change in $\hat{p}_t^{\rm mia}$ does not affect the rate of the primary hard scattering processes since – owing to the jet selection – the average parton transverse momentum is $\langle \hat{p}_t \rangle \approx 8$ GeV, which is well above the cut-off. Besides the effects of the underlying event energy, the jet cross section clearly depends on the rate of hard scattering processes.

Also shown in Fig. 4.35 is the ratio of the measured differential jet cross section (H1 data [68]) and the PYTHIA calculation with one hard parton scattering process per event. The data are compatible with the calculations using the GRV-LO parton distributions of the photon. The calculation, which uses the LAC1 parameterizations, gives a jet cross section which is too large.

Summary

1. Calculations of inclusive jet cross sections at large x_γ using quark distributions that were extracted from deep inelastic lepton–photon scattering experiments are compatible with the two–photon and photoproduction data.
2. The effect of the underlying event energy on the jet cross section is large: 40% at $E_t^{\rm jet} = 11$ GeV and $\eta^{\rm jet} = 2$. This has to be taken into account for meaningful conclusions on the low-x_γ parton distributions of the photon from jet data.
3. Comparisons of the jet data with calculations that describe the underlying event energy show that the jet data are sensitive to the parton density of the photon at small x_γ. The calculations, which use a moderately rising parton distribution function (GRV-LO), are compatible with the measured jet rate. A steeply rising parton density at low-x_γ (LAC1) is disfavored.

4.4.5 Measurement of the Gluon Distribution

Instead of adjusting the jet underlying event energy in the QCD calculations to the level observed in the data, the data can be corrected. This procedure has the advantage that the parton distributions of the photon can be directly obtained from the data.

In Fig. 4.28 it was shown that the observed large transverse energy is most weakly correlated with the jets coming from the primary parton scattering process. Therefore, the jet rate, which results from the large underlying event energy, can be subtracted on a statistical basis. In the following H1 analysis [64], the jet energies were corrected with respect to the underlying event energy $E_t^{\rm sia}$, which is expected from a single hard parton interaction per event. The jet energy correction $E_t^{\rm corr}$ was determined by a comparison of the transverse energy next to the jets in data, $E_t^{\rm data}$, and the QCD calculation of the PYTHIA generator. It was parameterized as a function of the jet rapidity $\eta^{\rm jet}$: $E_t^{\rm corr}(\eta^{\rm jet}) = E_t^{\rm data}(\eta^{\rm jet}) - E_t^{\rm sia}(\eta^{\rm jet})$, and amounted to $0.3 - 2.3$ GeV. The corrected jet energy $E_t^{\prime\rm jet} = E_t^{\rm jet} - E_t^{\rm corr}$ was required to be above the jet energy threshold $E_t^{\prime\rm jet} \geq 7$ GeV.

The di-jet event selection was based on an integrated luminosity of 0.29 pb^{-1} of the 1993 data-taking period, requiring the electron to be detected in the small angle electron detector. The kinematics correspond to quasi-real photon–proton scattering with $Q^2 \le 0.01\ \text{GeV}^2$ and $150 \le \sqrt{s_{\gamma p}} \le 250$ GeV. Both jets had rapidities between $0 \le \eta^{\text{jet}} \le 2.5$ and a rapidity distance between the two jets of $|\Delta\eta| \le 1.2$ in order to prevent the photon remnant from being misinterpreted as one of the hard jets.

The data were corrected to the parton fractional energy x_γ of the leading-order parton level (2.41). The parton energy was first estimated from the two jets with the highest transverse energy in the event using formula (4.2) and then corrected by an unfolding procedure [17]. The procedure requires a correlation between the parton x_γ and the reconstructed x_γ^{jets} as input. This was calculated using the QCD generator PYTHIA with multiple parton interactions. The energies of the calculated jets were corrected for the additional underlying event energy resulting from the multiple parton interactions. The unfolding procedure modifies the input parton x_γ distribution of the QCD calculation so as to produce agreement between the predicted jet distribution and that seen in the data. The procedure is independent of the input parton distribution, and is controlled by the comparison of other jet observables between the data and the modified calculation [64].

In Fig. 4.36, the two-jet distribution is shown as a function of the corrected fractional energy x_γ of the parton from the photon. The distribution shows two peaks. Following procedure 2) of Sect. 4.4.2, the peaks can be interpreted by a comparison with a leading order QCD calculation, separating the three different photon components, direct γ, quark and gluon contributions:

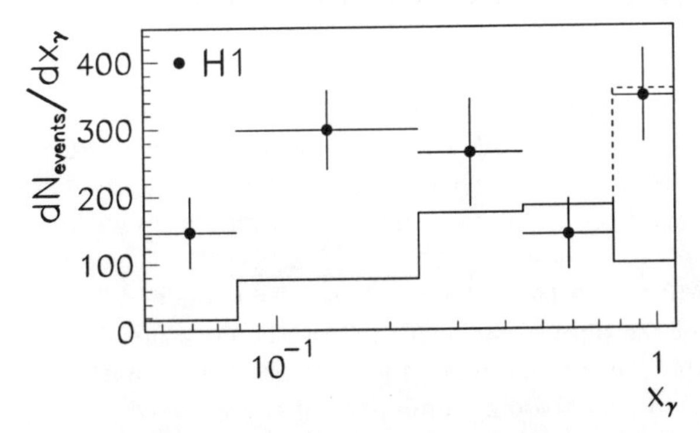

Fig. 4.36. The distribution of the fractional energy x_γ of the parton from the photon side is shown from H1 data. The *full histogram* is a QCD calculation of the PYTHIA generator using only the quark distribution function of the photon (GRV-LO parameterization [51]). The *dashed histogram* shows the direct photon contribution. At small parton momenta, the difference between the data and the calculations shows a significant gluon contribution of the photon (reprinted from [64] with kind permission from Springer Verlag)

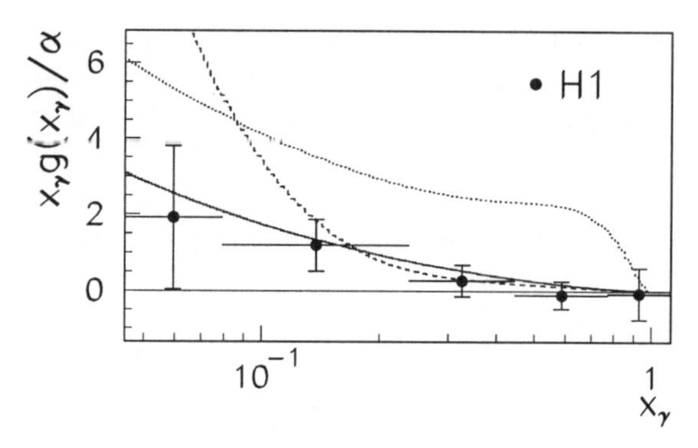

Fig. 4.37. The measured gluon distribution of the photon $x_\gamma g(x_\gamma)$ divided by α is shown from H1 data. The average scale is $\langle p_t^2 \rangle = 75\ \mathrm{GeV}^2$. The curves represent parameterizations of the gluon distribution using the results of deep inelastic eγ scattering experiments (*full*=GRV [51], *dashed*=LAC1, *dotted*=LAC3 [1]) (reprinted from [64] with kind permission from Springer Verlag)

1. The *resolved* photon–proton interactions with a *quark* on the photon side at the parton scattering process are shown as the full histogram in Fig. 4.36. The matrix elements of the parton scattering are predicted by QCD, whereas the parton densities of the photon and the proton have to be taken from deep inelastic eγ, or ep scattering experiments. Here the GRV-LO parameterizations were used [49, 51]). Around $x_\gamma = 0.5$, this component alone is compatible with the jet data observed in the photoproduction events. Essentially, in this kinematic range, the photon splits into a quark and an anti-quark, each carrying half of the photon energy.
2. The matrix elements of *direct* photon–proton scattering are predicted by QCD theory. The dashed histogram above $x_\gamma \geq 0.77$ represents a LO QCD calculation using the GRV-LO parton distribution functions for the proton. When the direct photon component is added to the resolved γ processes with a quark from the photon side, the result agrees with the data.
3. Below $x_\gamma \leq 0.2$ the observed two-jet rate cannot be explained by the quark contributions of the photon, as measured by the deep inelastic eγ scattering experiments. This observed large jet rate above the quark contribution exhibits a significant *gluon* component of the *resolved* photon. Comparison with a calculated two-jet distribution including gluons in the photon gave the measured gluon distribution in the photon shown in Fig. 4.37.

The error bars of the gluon measurement in Fig. 4.37 result from the uncertainty in the knowledge of the absolute calorimeter energy scale, from the uncertainty in the determination of the underlying event energy for the jet energy correction, and from the uncertainty in the quark distribution of the photon.

Three leading-order parameterizations of the gluon distribution of the photon are shown in the same figure (LAC1 = dashed, LAC3 = dotted, GRV-LO = full [1, 51]). In spite of the sizable error bars, the measurement gives new information: the gluon distribution is neither very large at high x_γ (LAC3), nor steeply rising towards small values of x_γ (LAC1), but shows a shape and size that is, within error bars, reproduced by the GRV-LO parameterization.

Summary

For the first time, the gluon distribution of the photon was determined from data at the leading-order level down to low parton fractional energies $x_\gamma \geq 0.04$ at the large average scale of $\langle p_t^2 \rangle = 75\ \text{GeV}^2$.

4.4.6 Double Differential Di-Jet Cross Section

So far, the parton distributions of the photon were studied using differential jet and particle cross sections as a function of one observable, for example the jet rapidity η or the parton fractional energy x_γ. Here the double differential di-jet cross section is analyzed in terms of the two jet observables x_γ^{jets} and E_t^{jet} which correspond to the relevant variables of the parton distributions of the photon: x_γ and the renormalization scale μ_{fac}. Comparisons between the measured and the calculated di-jet cross section give much stronger constraints on the parton distributions of the photon than the single differential distributions.

The following H1 analysis is based on 1994 data with an integrated luminosity of 2 pb^{-1} [46]. The scattered electron remained unmeasured (untagged events). The virtuality of the photon is, therefore, at most $Q^2 = 4\ \text{GeV}^2$ where 80% of the events are below $Q^2 = 0.1\ \text{GeV}^2$. The photon energy was reconstructed from the hadronic final state (2.39) and is – after detector corrections – between $0.2 < y < 0.83$. The jets were found with a cone algorithm requiring at least $E_t^{\text{jet}} = 8$ GeV collected in a cone of size $R = 0.7$. The jet rapidities were restricted to $-0.5 < \eta^{\text{jet}} < 2.5$. The rapidity difference between the two jets with the highest E_t^{jet} in the event had to be below $|\Delta\eta^{\text{jet}}| < 1$. The latter cut corresponds to a restriction of the parton scattering angle $0 < \cos\hat{\theta} < 0.46$.

The parton fractional energy x_γ^{jets} was determined with the two jets with the highest E_t^{jet} in the event using (4.3). The scale E_t was chosen to be the transverse energy of the jet with the second largest E_t^{jet}.

In Fig. 4.38 the double differential di-jet cross section is shown as a function of E_t in bins of x_γ^{jet}. The data were corrected for detector effects. Here no corrections for the underlying event energy were applied. The inner error bars reflect the statistical errors, the outer error bars represent the quadratic sum of the statistical and systematic errors. The dominant systematic error results from the uncertainty in the knowledge of the calorimeter energy scale.

In the same figure, the full curves represent the calculation of the PYTHIA generator with multiple parton interactions. For the parton distributions of the proton and the photon, the GRV-LO parameterizations were used [49, 51]. The calculation gives an overall satisfactory description of the data. It falls low in the region of high parton fractional energies $x_\gamma^{\rm jets}$ and large scales E_t. This indicates the regions where modifications to the input parton distribution of the photon are needed.

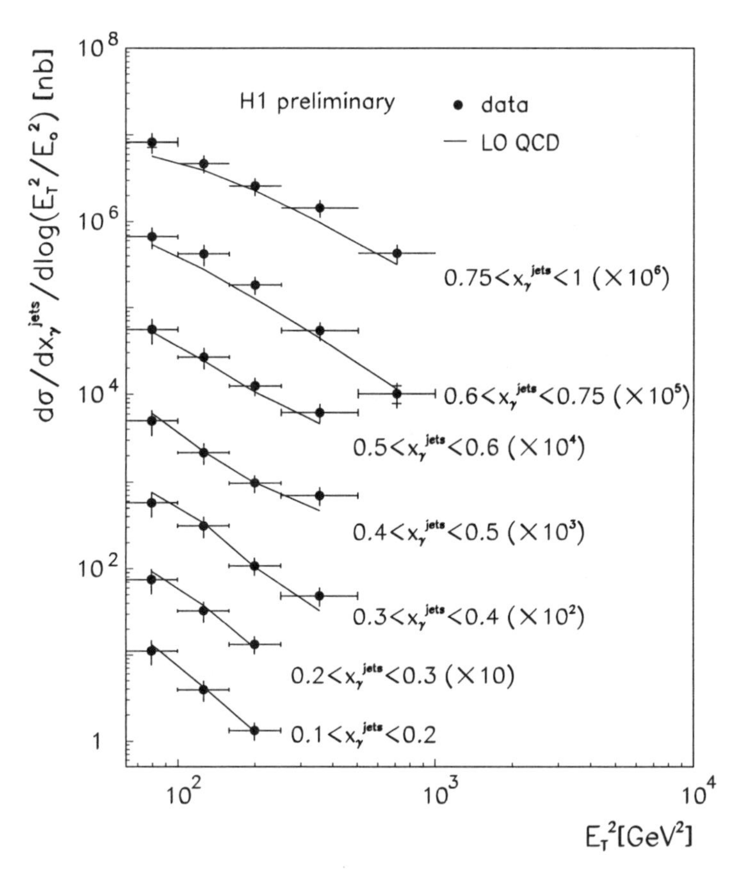

Fig. 4.38. The measured double differential di-jet ep cross section is shown as a function of the logarithm of the squared jet transverse energy E_t^2 in bins of the reconstructed parton fractional energy $x_\gamma^{\rm jets}$ (*full circles*, $E_\circ^2 = 1$ GeV^2). The cross section is integrated over the photon virtuality $Q^2 < 4\,{\rm GeV}^2$, the relative photon energy $0.2 < y < 0.83$ and the difference of the jet pseudo-rapidities $|\Delta\eta^{\rm jets}| < 1$. The data are compared with the QCD calculation of the PYTHIA generator with multiple parton interactions using the GRV-LO parton distribution functions for the proton and the photon (*full curves* [49, 51])

Summary

The measured double differential di-jet cross section constrains the parton distributions of the photon with the large number of data points in the ranges of the parton fractional energy $0.1 < x_\gamma^{\text{jets}} < 1$ and the scale $63 < E_t^2 < 1000\ \text{GeV}^2$.

4.4.7 Measurement of the Effective Parton Distribution

As explained in Sect. 2.1.3, perturbative QCD predicts a unique characteristic of the quark density $f_{\text{q}/\gamma}$ in the photon (2.18): $f_{\text{q}/\gamma}$ increases at large parton fractional energies x_γ and depends directly on the factorization scale:

$$f_{\text{q}/\gamma} \propto \ln \mu_{\text{fac}}^2 \ . \tag{4.12}$$

The scale dependence reflects the anomalous component of the photon and results from the pointlike coupling of the photon to quark–anti-quark pairs. In photon–proton processes the scale μ_{fac}^2 corresponds to the transverse momentum of the scattered partons $\mu_{\text{fac}}^2 = \xi\, p_t^2$ with $\xi > 0$.

The aim of this H1 analysis is to test the logarithmic scale dependence of (4.12). In the jet data, only the combined quark and gluon densities are measured. We follow here the third method, explained in Sect. 4.4.2, and extract the effective parton distribution of the photon $\tilde{f}_\gamma = \sum_{\text{n}_\text{f}} (\text{q}_\gamma + \overline{\text{q}}_\gamma) + (9/4)\, \text{g}_\gamma$ from the di-jet data in Fig. 4.38. A kinematic region around $x_\gamma \approx 0.5$ is selected where the quark contribution dominates the gluon in the photon and the direct γp processes do not contribute.

For the extraction of $\tilde{f}_\gamma(x_\gamma, p_t^2)$ two steps are needed: first the double differential di-jet cross section $\text{d}^2\sigma/\text{d}x_\gamma^{\text{jets}}/\text{d}\log{(E_t/E_\circ)}^2$ is corrected to the level of the di-parton cross section $\text{d}^2\sigma/\text{d}x_\gamma/\text{d}\log{(p_t/p_\circ)}^2$, where x_γ gives the energy fraction at the leading order parton level, and p_t denotes the parton transverse momentum of the two final-state partons. A two-dimensional unfolding procedure [36] was used for the corrections together with the PYTHIA generator with multiple parton interactions. The corrections concern higher-order QCD effects, the influence of the underlying event energy and fragmentation effects. In the second step, the comparison between the measured and calculated di-parton cross sections gives $\tilde{f}_\gamma$ of the data.

The measured effective parton distribution of the photon $\tilde{f}_\gamma$ is shown in Fig. 4.39 as a function of the transverse parton momentum p_t. The data were averaged in the range $0.4 < x_\gamma < 0.7$ and normalized by the fine structure constant α. The error bars reflect the quadratic sum of the statistical and systematic errors. The dominant systematic errors result from the uncertainty in the knowledge of the calorimeter energy scale and the uncertainty in the treatment of the underlying event effects. The measured $\tilde{f}_\gamma$ increases with increasing p_t at a rate that is compatible with the logarithmic increase predicted by QCD theory.

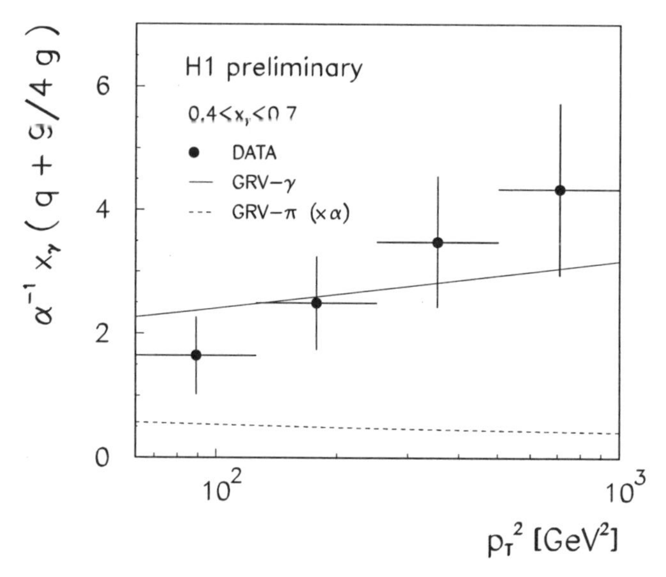

Fig. 4.39. The leading-order effective parton distribution of the photon $x_\gamma \tilde{f}_\gamma$ is shown as a function of the squared parton transverse momentum p_t^2 (*circles*: H1 data). Here q represents the sum over all quarks and anti-quarks, and g denotes the gluon density. The distribution was normalized to the fine structure constant α. The data were averaged over the interval of the parton fractional energies $0.4 < x_\gamma < 0.7$. The data are compared to the effective parton distributions of the photon (*full curve*) and the pion – omitting the factor $1/\alpha$ – (*dashed curve*) as derived from the GRV-LO parton distribution functions [50, 51]

In the same Fig. 4.39, the data are compared with the effective parton distribution as derived from the GRV-LO parameterization of the parton distribution in the photon (full curve) [51]. The summed quark densities give here 80% of the total $\tilde{f}_\gamma$, independent of p_t. The parameterization is compatible with the result from the jet data. Also shown is the effective parton distribution of the pion using the GRV-LO parameterization [50]. Here the normalization to α has been omitted. The shape of the data is clearly distinct from the scale dependence of a purely hadronic parton distribution.

The QCD prediction of the anomalous photon component has been confirmed in measurements of the photon structure function F_2^γ before (Fig. 3.3). The precision of the γp data in Fig. 4.39 is very competitive with the precision of the data from deep inelastic eγ scattering. The observed scale dependencies are compatible within errors. If we identify the factorization scale of the jet data with $\mu_{\mathrm{fac}}^2 = p_t^2$, and in the structure function measurements with $\mu_{\mathrm{fac}}^2 = Q^2$, then the γp data test the scale dependence in a new kinematical region up to $\mu_{\mathrm{fac}}^2 = 10^3$ GeV^2. Depending on the future luminosity at HERA, the region can be extended up to $\mu_{\mathrm{fac}}^2 = 10^4$ GeV^2.

Summary

1. When we use the quark distributions resulting from deep inelastic lepton–photon scattering experiments in the QCD calculation of inclusive jet cross sections at large x_γ we obtain results that are compatible with the two–photon and photoproduction data.
2. For the first time, the gluon distribution of the photon has been extracted from di-jet data for parton fractional energies $x_\gamma \geq 0.04$. The gluon component is not large at high x_γ. It also does not rise steeply towards low x_γ (e.g., LAC1 parameterization), but shows a moderate rise (e.g., GRV-LO). Studies of the inclusive jet and particle cross sections in γp and $\gamma\gamma$ experiments confirm this measurement.
3. For the first time, an effective parton distribution was extracted from the analysis of double differential di-jet cross sections. This jet measurement confirms the basic QCD prediction of the anomalous photon component with the same precision as obtained in deep inelastic lepton–photon scattering experiments.
4. Comparisons of measured jet cross sections with next-to-leading QCD calculations at small $x_\gamma < 0.4$ are difficult, because the large observed underlying event energy is not included in the calculations, but has a large influence on the results. Comparisons of particle cross section measurements with NLO QCD calculations will provide a consistent way of extracting NLO quark and gluon distributions of the photon at small x_γ: these distributions are insensitive to the underlying event energy effects at sufficiently high hadron transverse momenta $p_t > 3$ GeV.

5. Status and Future

5.1 Status: Hard Photoproduction at HERA

The HERA machine together with the experiments H1 and ZEUS has proved to be an important facility for the understanding of the photon. The experiments have already provided new information on the quasi-real photon, complementary and beyond the results obtained in photon–nucleon and two-photon experiments. All photoproduction measurements at HERA with large transverse energy E_t in the hadronic final state are compatible with the QCD picture of hard parton–parton scattering processes between the photon and the proton.

The presence of hard scattering in photoproduction processes at HERA was established by the following observations:

1. Particles with large transverse momentum p_t: the inclusive charged particle cross sections were measured in terms of the rapidity and transverse momentum up to $p_t = 12$ GeV.
2. Jet production: differential jet cross sections were measured as functions of the jet rapidity and transverse jet energy up to $E_t^{\text{jet}} = 40$ GeV.

The shapes of jets from γp collisions are very similar to those found in $\overline{\text{p}}$p and $\gamma\gamma$ scattering. They show narrowing of the jet width with increasing transverse jet energy. The photoproduction jets and jets from $\overline{\text{p}}$p and $\gamma\gamma$ collisions are related by a simple scaling law on the beam energies.

In the central rapidity region of the γp collision the underlying event energy next to the γp produced jets exceeds by far that expected from QCD calculations of a single hard parton–parton scattering process per event. Since the E_t^{jet} spectrum is falling so steeply, differences in the underlying event energy have a large influence on the jet cross sections. To gain information on the partonic scattering process from a comparison between the measured and calculated inclusive jet cross sections, the underlying event energy has to be taken into account.

To understand the relation between the measured jet cross sections and the underlying event, detailed studies of the transverse energy flow at large E_t in the photoproduction events were made, with special emphasis on the additional energy seen in the data:

1. The differential transverse energy cross section shows a high E_t tail, as expected from hard scattering processes.
2. The maximum transverse energy flow is observed in the central rapidity region of the γp collision ($\eta_{\gamma p} = 0$).
3. The energy depositions in the central γp collision region ($\eta_{\gamma p} = 0$) are weakly correlated with the other energy depositions in the event.
4. The transverse energy density at $\eta_{\gamma p} = 0$ outside of jets increases with decreasing parton fractional energies x_γ from the level of deep inelastic ep scattering events (direct γp processes) to the level observed in jet events of $\overline{\text{p}}$p collisions.

These distributions are not described by models that calculate the exclusive final state from one hard parton–parton scattering process per event. A natural explanation for the observed transverse energy distributions are multiple parton interactions with on average up to three interactions per event. The γp results complement the measurement of double parton scattering processes, reported by $\overline{\text{p}}$p experiments.

Basic QCD predictions were confirmed by the data:

1. The direct interactions of quasi-real photons with protons have been confirmed by studying the 2-jet distribution of the fractional energy x_γ of the parton from the photon.
2. The presence of resolved photon–proton processes was shown by the observation of the photon remnant.
3. The distributions of the parton scattering angle in direct and resolved photon processes were found to be different, and are correctly predicted by QCD.
4. The presence of higher order QCD effects was shown by the transverse energy imbalance in di-jet events and by the observation of multi-jet events.

Distributions of the partons from the photon were analyzed:

1. The distribution of the intrinsic transverse momentum k_t^γ of the partons from the photon was found to be harder than that of hadrons.
2. Measurements of the single inclusive jet cross sections and the di-jet cross sections at large x_γ provide access to the extraction of information on the quark distributions of the photon at $x_\gamma \approx 1$.
3. At parton fractional energies of about $x_\gamma = 0.5$, the measurement of double differential di-jet cross sections allowed for the first extraction of an effective parton distribution in the photon. The data confirm the logarithmic scale dependence of the parton distributions, which is predicted by QCD theory owing to the pointlike splitting of the photon into a quark–anti-quark pair.
4. At small x_γ, measurements of di-jet production resulted in the first measurement of the leading order gluon distribution in the photon in the range $0.04 < x_\gamma < 1$ at a scale $\langle p_t^2 \rangle = 75$ GeV^2.

The results reviewed in this book can be understood as a first successful round of high-E_t photoproduction analyses at HERA. Both experiments have delivered complementary results according to the strength of their detectors: the H1 experiment used the fine granularity calorimeter in the central rapidity region of the γp collision to study the effects of multiple parton interactions and the small-x_γ parton content of the photon. The ZEUS experiment used the uniform coverage of the photon fragmentation region to study the photon remnant and direct photon interactions. Both experiments analyzed the resolved photon processes at large parton fractional energies $x_\gamma \approx 1$.

5.2 Prospects of High Energy Photon Interactions

Since 1996, at LEP, electrons and positrons have been collided at center-of-mass energies $\sqrt{s_{ee}} = 130$–190 GeV (LEP2). Two-photon physics will be studied at center-of-mass energies up to $\sqrt{s_{\gamma\gamma}} = 100$ GeV [8]. The measurements of F_2^γ will then cover the large kinematical region $10^{-4} < x_\gamma < 1$ at scales between $3 < Q^2 < 10^3$ GeV^2. Within the next few years, the accuracy of these measurements is expected to be at the level of a few percent. The gluon will be studied in jet events in the region around $x_\gamma = 0.1$ at scales above $p_t^2 = 25$ GeV^2. Charm production may extend the region down to $x_\gamma = 10^{-3}$. New information will be obtained on the parton content of virtual photons. LEP2 also gives access to detailed measurements of the triple boson vertex γWW [57].

At HERA the integrated luminosity is expected to increase by 1–2 orders of magnitude within the next few years [9]. Cross section measurements with a statistical precision of a few percent are possible, for example, for charged particles at high transverse momenta and for jets. An improved understanding of the detector effects is being achieved for such precision measurements. These data will allow the extraction of accurate NLO parton distributions of the photon (e.g., [16]).

Studies of the parton distributions of virtual photons have recently been started [138] and are expected to shed new light on the transition region between photoproduction and deep inelastic lepton–proton scattering. Furthermore, the analysis of diffractive hard scattering processes revealed a parton structure which can be assigned to the Pomeron (e.g., [65]). With the knowledge of the parton content in the photon already achieved, diffractive hard photoproduction processes provide a unique way of studying the gluon content of the Pomeron (e.g., [139]).

The next generation photon facility is already a very active field of experimental research and theoretical studies [90]: linear colliders will operate in ee, eγ, or $\gamma\gamma$ collision modes at center-of-mass energies between top–anti-top quark production and $\sqrt{s} = 2000$ GeV. They will act as clean W factories, and provide access to the very low x_γ parton content of the photon at high Q^2. If the Higgs boson is found in the few hundred GeV region, then the photon collider is an ideal place to measure its properties in order to confirm the standard model, or to find new physics.

References

1. H. Abramowicz, K. Charchula, A. Levy, Phys. Lett. B269 (1991) 458
2. AFS Collab., T. Akesson et al., Phys. Lett. B128 (1983) 354
3. AFS Collab., T. Akesson et al., Z. Phys. C34 (1987) 163
4. AMY Collab. B .J. Kim et al., Phys. Lett. B325 (1994) 248
5. P. Aurenche et al., Phys. Rev. D45 (1992) 92
6. P. Aurenche et al., Z. Phys. C56 (1992) 589
7. P. Aurenche, M. Fontannaz and J .P. Guillet, Phys. Lett. B338 (1994) 98
8. P. Aurenche, G .A. Schuler et al., '$\gamma\gamma$ Physics', in physics at LEP2, eds. G. Altarelli, T. Sjöstrand and F. Zwirner, CERN 96-01 (yellow report), HEP-PH-9601317 (1996)
9. W. Bartel et al., Proc. of the Workshop 'Future physics at HERA', eds. G. Ingelman, A. DeRoeck, R. Klanner, Hamburg (1996) 1095
10. T .H. Bauer et al., Rev. Mod. Phys. 50 (1978) 261, Rev. Mod. Phys. 51 (1979) 407
11. A .C. Bawa, W .J. Stirling, J. Phys. G15 (1989) 1339
12. Ch. Berger, PITHA 83/22 (1983), 'Int. Symp. on Lepton Photon Interactions', Ithaca, N.Y., Publ. in Lepton/Photon Symp. (1983) 376
13. A. Baldini et al., Landolt-Börnstein's New Series I/12B (1988) 345
14. Ch. Berger and W. Wagner, Phys. Rep. 146 (1987) 1
15. J. Binnewies, B .A. Kniehl and G. Kramer, Phys. Rev. D52 (1995) 4947
16. J. Binnewies, M. Erdmann, B .A. Kniehl and G. Kramer, Proc. of the Workshop 'Future physics at HERA', eds. G. Ingelman, A. DeRoeck, R. Klanner, Hamburg (1996) 549
17. V. Blobel, DESY 84-118 (1984), Proc. of the 1984 CERN School of Computing, Aiguablava (Spain), CERN (1985)
18. D. Bödeker, G. Kramer, S .G. Salesch, Z. Phys. C63 (1994) 471
19. F .W. Bopp et al., Phys. Rev. D49 (1994) 3236
20. F .W. Bopp et al., Comp. Phys. Commun. 83 (1994) 107
21. S .J. Brodsky, et.al, Phys. Rev. D19 (1979) 1418
22. BS Collab., B. Alper et al., Nucl. Phys. B100 (1975) 237
23. V .M. Budnev et al., Phys. Rep. 15 (1975) 181
24. A. Buniatian, PhD thesis, Hamburg, DESY internal report FH1K-95-04 (1995)
25. J .M. Butterworth et al., CERN-TH-95-83 (1995), Proc. of the Conference 'PHOTON 95', Sheffield, England (1995) 175
26. J .M. Butterworth, L. Feld, M. Klasen, G. Kramer, Proc. of the Workshop 'Future physics at HERA', eds. G. Ingelman, A. DeRoeck, R. Klanner, Hamburg (1996) 554
27. A. Capella et al., Phys. Rev. Lett. 58 (1987) 2015
28. A. Capella et al., Phys. Rep. 236 (1994) 227
29. P. Capiluppi et al., Nucl. Phys. B70 (1974) 1
30. S. Catani, Yu .L. Dokshitser and B .R. Webber, Phys. Lett. B285 (1992) 291

31. CDF Collab., F. Abe et al., Phys. Rev. D45 (1992) 2249
32. CDF Collab., F. Abe et al., Phys. Rev. D47 (1993) 4857
33. M. Chen and P. Zerwas, Phys. Rev. D12 (1975) 187
34. J. Chyla, Phys. Lett. B320 (1994) 186
35. B .L. Combridge and C .J. Maxwell, Nucl. Phys. B239 (1984) 429
36. G. D'Agostini, Nucl. Instrum. Meth. A362 (1995) 487
37. DELPHI Collab., P. Abreu et al., Z. Phys. C69 (1996) 223
38. A. Donnachie, P .V. Landshoff, Phys. Lett. B296 (1992) 227
39. M. Drees and K. Grassie, Z. Phys. C28 (1985) 451
40. M. Drees, MAD-PH-797, Proc. of the '23rd Intern. Symp. Ultra-High Energy Multiparticle Phenomena', Aspen, Colorado (1993)
41. D .W. Duke and J .F.Owens, Phys. Rev. D26 (1982) 1600
42. E683 Collab., D. Adams et al., Phys. Rev. Lett. 72 (1994) 2337
43. R. Engel, Z. Phys. C66 (1995) 203
44. R. Engel and J. Ranft, Phys. Rev. D54 (1996) 4244
45. R. Engel et al., Phys. Rev. D46 (1992) 5192
46. M. Erdmann and H. Rick, Proc. of the 28th International Conference on High Energy Physics, Warsaw, Poland (1996)
47. J .H. Field, F. Kapusta, L. Poggioli Phys. Lett. 181B (1986) 362
48. J .H. Field, F. Kapusta, L. Poggioli Z. Phys. C 36 (1987) 121
49. M. Glück, E. Reya and A. Vogt, Z. Phys. C53 (1992) 127
50. M. Glück, E. Reya and A. Vogt, Z. Phys. C53 (1992) 651
51. M. Glück, E. Reya and A. Vogt, Phys. Rev. D46 (1992) 1973
52. M. Glück, E. Reya, M. Stratmann, Phys. Rev. D51 (1995) 3220
53. L .E. Gordon and J. K. Storrow, Z. Phys. C56 (1992) 307
54. L. E. Gordon and J. K. Storrow, ANL-HEP-PR-96-33, HEP-PH-9607370 (1996)
55. L. E. Gordon, J. K. Storrow, Phys. Lett. B291 (1994) 320
56. L. E. Gordon and W. Vogelsang, Phys. Rev. D52 (1995) 58
57. G. Gounaris, J. L. Kneur, D. Zeppenfeld et al., 'Triple Gauge Boson Couplings', in physics at LEP2, eds. G. Altarelli, T. Sjöstrand and F. Zwirner, HEP-PH-9601233 (1996)
58. M. Greco, A. Vicini, Nucl. Phys. B415 (1994) 386
59. V. N. Gribov and L. N. Lipatov, Yad. Fiz. 15 (1971) 781, (Sov. J. Nucl. Phys. 15 (1972) 438)
60. H1 Collab., T. Ahmed et al., Phys. Lett. B297 (1992) 205
61. H1 Collab., I. Abt et al., Phys. Lett. B314 (1993) 436
62. H1 Collab., I. Abt et al., Phys. Lett. B328 (1994) 176
63. H1 Collab., T. Ahmed et al. Nucl. Phys. B435 (1995) 3
64. H1 Collab., T. Ahmed et al., Nucl. Phys. B445 (1995) 195
65. H1 Collab., T. Ahmed et al., Phys. Lett. B348 (1995) 681
66. H1 Collab., S. Aid et al., Phys. Lett. B358 (1995) 412
67. H1 Collab., S. Aid et al., Z. Phys. C69 (1995) 27
68. H1 Collab., S. Aid et al., Z. Phys. C70 (1996) 17
69. H1 Collab., I. Abt et al., internal report DESY-96-01 (1996)
70. K. Hagiwara, T. Izubuchi, M. Tanaka and I. Watanabe, Phys. Rev. D51 (1995) 3197
71. W. Hoprich, diploma thesis, Heidelberg (1995)
72. P. Hoyer, NORDITA-95-65-P (1995), Workshop on 'Deep Inelastic Scattering and QCD', Paris, France (1995)
73. J. E. Huth et al., Fermilab-Conf-90/249-E (1990)
74. H. Hufnagel, PhD thesis, Heidelberg (1994)

75. M. Jacob, Proc. XVI Int. Conference on High Energy Physics, eds. J.D. Jackson and A. Roberts, Chicago, Vol.3 (1972) 373
76. F. Kapusta, Z. Phys. C42 (1989) 225
77. M. Klasen, G. Kramer and S. G. Salesch, Z. Phys. C68 (1995) 113
78. M. Klasen, G. Kramer, DESY-96-246, HEP-PH-9611450 (1996)
79. B. A. Kniehl and G. Kramer, Z. Phys. C62 (1994) 53
80. H. Kolanoski, 'Two Photon Physics at e^+e^- Storage Rings', Springer Tracts in Mod. Phys. 105, Heidelberg: Springer (1984) p.187
81. H. Kolanoski and P. Zerwas, DESY-87-175 (1987), 'High Energy Electron-Positron Physics', Ed. A. Ali and P. Söding, Singapore, Singapore: World Scientific (1988)
82. G. Kramer, S. G. Salesch, Z. Phys. C61 (1994) 277
83. G. Kramer and S. G. Salesch, Phys. Lett. B333 (1994) 519
84. A. Levy, 'Photoproduction at HERA', DESY Academic Training Program (1992)
85. G. Marchesini and B. R. Webber, Nucl. Phys. B238 (1984) 1
86. G. Marchesini and B. R. Webber, Nucl. Phys. B310 (1988) 461
87. G. Marchesini and B. R. Webber, Phys. Rev. D38 (1988) 3419
88. G. Marchesini et al., Comput. Phys. Commun. 67 (1992) 465
89. A. D. Martin, W. J. Stirling and R. G. Roberts, Phys. Rev. D47 (1993) 867
90. D. J. Miller, DESY 95-183 (1995)
91. NA14 Collaboration, E. Auge et al., Phys. Lett. 168B (1986) 163
92. OMEGA γ Collab., R. J. Apsimon et al., Z. Phys. C43 (1989) 63
93. H. Baer, J. Ohnemus and J. F. Owens, Phys. Rev. D40 (1989) 2844
94. E. Paul, DESY-92-026 (1992), 'Multiparticle Dynamics', Wuhan, China (1991) 267
95. P. Pfeiffenschneider, diploma thesis, Aachen (1994)
96. H. Plothow-Besch, Comput. Phys. Commun. 75 (1993) 396
97. H. Rick, Proc. of the Workshop 'Deep Inelastic Scattering and QCD', ed. G.D'Agostini, Rome, Italy (1996)
98. H. Rick, PhD thesis, in preparation (1997)
99. J. J.Sakurai, Ann. Phys. 11 (1960) 1
100. S. G. Salesch, DESY-93-196 (1993)
101. S. G. Salesch, Manual for NLO QCD program 'JETSAM', H1 internal note (1995)
102. G. A. Schuler and T. Sjöstrand, Nucl. Phys. B407 (1993) 539
103. G. A. Schuler and T. Sjöstrand, CERN-TH-7193-94 (1994), Proc. of the Conference 'Two-Photon Physics from DAPHNE to LEP200 and Beyond', ed. F. Kapusta and J. Parisi, Singapore: World Scientific, Paris (1994)
104. G. A. Schuler and T. Sjöstrand, Z. Phys. C68 (1995) 607
105. G. A. Schuler and T. Sjöstrand, Phys. Lett. B376 (1996) 193
106. S. Söldner-Rembold, Proc. of the 28th International Conference on High Energy Physics, Warsaw, Poland (1996)
107. C. Schwanenberger, diploma thesis, Heidelberg (1995)
108. T. Sjöstrand, M. Bengtsson, Comput. Phys. Commun. 43 (1987) 367
109. T. Sjöstrand and M. van Zijl, Phys. Rev. D36 (1987) 2019
110. T. Sjöstrand, CERN-TH-6488 (1992), Comput. Phys. Commun. 82 (1994) 74
111. T. Sjöstrand, J. K. Storrow, A. Vogt, J. Phys. G22 (1996) 893
112. M. Steenbock, PhD thesis, Hamburg, DESY-F11-F22-96-01 (1996)
113. TOPAZ Collab., H. Hayashii et al., Phys. Lett. B314 (1993) 149
114. TOPAZ Collab., K. Muramatsu et al. Phys. Lett. B332 (1994) 477
115. TOPAZ Collaboration, R. Enomoto et al., Phys. Rev. D50 (1994) 1879
116. TPC/2γ Collab., M. P. Cain et al. Phys. Lett. 147B (1984) 232

117. TPC/2γ Collab., H. Aihara, et. al, Z. Phys. C34 (1987) 1, Phys. Rev. Lett. 58 (1987) 97
118. UA1 Collab., G. Arnison et al., Phys. Lett. B132 (1983) 214
119. UA1 Collab., C. Albajar et al., Z. Phys. C36 (1987) 33
120. UA1 Collab., C. Albajar et al., Nucl. Phys. B309 (1988) 405
121. UA1 Collab., C. Albajar et al., Nucl. Phys. B335 (1990) 261
122. UA2 Collab., J.Alitti et.al. Phys. Lett. B268 (1991) 145
123. A. Vogt, LMU-11-94 (1994), Proc. of Workshop 'Two-Photon Physics at LEP and HERA', Ed. G. Jarlskog and L. Jönnson, ISBN 91-630-2886-7, Lund, Sweden (1994) 141
124. OPAL Collab., J. Ward, Proc. of the Conference 'Two-Photon Physics from DAPHNE to LEP200 and Beyond', ed. F. Kapusta and J. Parisi, Singapore: World Scientific, Paris (1994)
125. B. R. Webber, Nucl. Phys. B238 (1984) 492
126. C. F. von Weizsäcker, Z. Phys. 88 (1934) 612
127. E. J. Williams, Kgl. Danske Vidensk. Selskab. Mat.-Fiz. Medd. 13 N4 (1935)
128. E. Witten, Nucl. Phys. B120 (1977) 189
129. Y. Yamazaki, Proc. of the 28th International Conference on High Energy Physics, Warsaw, Poland (1996)
130. ZEUS Collab., M. Derrick et al. Phys. Lett. B297 (1992) 404
131. ZEUS Collab., ZEUS internal note (1993)
132. ZEUS Collab., M. Derrick et al., Z. Phys. C63 (1994) 391
133. ZEUS Collab., M. Derrick et al., Phys. Lett. B322 (1994) 287
134. ZEUS Collab., M. Derrick et al., Phys. Lett. B346 (1995) 399
135. ZEUS Collab., M. Derrick et al., Phys. Lett. B348 (1995) 665
136. ZEUS Collab., M. Derrick et al., Phys. Lett. B354 (1995) 163
137. ZEUS Collab., M. Derrick et al., Z. Phys. C67 (1995) 227
138. ZEUS Collab., M. Derrick et al., Int. Europhys. Conference on High Energy Physics, contributed paper EPS-0384, Brussels (1995)
139. ZEUS Collab., M. Derrick et al., Phys. Lett. B356 (1995) 129
140. ZEUS Collab., M. Derrick et al., Phys. Lett. B384 (1996) 401

Abbreviations and Variables

List of Abbreviations

$\boldsymbol{a}$	four vector
$\mathrm{\bar{a}}$	anti-particle
$\overline{a}]$	average of two variables
$\langle a \rangle$	averaged value
$\hat{a}$	variable in the parton–parton center-of-mass system
a_{ij}	variable in the center-of-mass system of particles i and j
cal	calorimeter
CM	center-of-mass
CMS	center-of-mass system
corr	correction
DIS	deep inelastic scattering
dp	double parton scattering
eff	effective
ev	event
fac	factorization
frag	fragmentation
ia	interaction
LO	leading order
max	maximum
meas	measured
mia	multiple parton scattering in one event
min	minimum
miss	missing
NLO	next-to-leading order
nondiff	nondiffractive
O()	order of
QCD	quantum chromodynamics
QED	quantum electrodynamics
ren	renormalization
SES	single effective subprocess
sia	single parton interaction in one event
sp	spectator partons
VDM	vector meson dominance
vis	visible

List of Variables

α	fine structure constant
α_s	strong coupling constant
γ	photon
γ^*	virtual photon
γ_{V}	coupling constant of photon to vector mesons
Δ	difference
η	pseudo-rapidity
φ	azimuthal direction
Λ_{QCD}	QCD parameter
μ	scale parameter (fac=factorization scale, ren=renormalization scale, frag=fragmentation scale)
π	π meson
ϱ	ϱ meson
σ	cross section
τ	target particle
θ	polar angle
$D_{\mathrm{h}/i}$	fragmentation function: hadron h from parton i
e	electron
e′	scattered electron
e	electric charge
E	energy
E_t	transverse energy
$f_{i/j}$	parton density: parton i in particle j
$\tilde{f}_j$	effective parton density: parton in particle j
F_2	structure function
g	gluon
h	hadron
k_t	intrinsic transverse momentum
L	luminosity
m	mass
M	matrix element
N	nucleon
p	proton
p	momentum
p_t	transverse momentum
P_{ij}	Altarelli-Parisi splitting functions: parton j radiates parton i
q	quark
Q^2	photon virtuality
R	ratio
R	radius in (η, φ) space
s	squared center-of-mass energy
t	squared momentum transfer
u	squared momentum transfer

V vector meson
W W boson
x parton fractional energy
X hadronic final state
y scaled photon energy
z fractional energy

Index

deep inelastic scattering
- ep 6
- ep 39, 78
- eγ 29

detector
- H1 42
- ZEUS 42

energy
- correlations 76
- differential cross section 72
- flow 40, 75

equivalent photon approximation 8

final-state variables: $p_t, \varphi, \theta, \eta$ 38
fragmentation
- γ fragmentation region 40
- p fragmentation region 40
- function 17

Heisenberg uncertainty relation 6
HERA 41
HERWIG 27

jet
- algorithms 53
- configurations 39
- cross section 16, 21, 58, 86
- pedestal 54
- photon remnant 61
- profile 55, 56, 60, 62
- width 57

luminosity
- integrated 42
- measurement 46

mid-rapidity 40
minimum bias data 75, 78, 80

particle
- cross sections 16, 21, 49, 84

parton
- center-of-mass energy $\hat{s}$ 13, 18, 21
- cross section 13, 21
- distributions in the γ 32, 82
- distributions in the γ 19
- double parton scattering 71
- effective parton distribution 19, 84
- fractional energy x 20
- kinematics 20, 39
- matrix elements 13
- multiple parton scattering 22, 26, 27, 56, 70, 86, 90
- processes 14
- scattering angle $\hat{\theta}$ 13, 20, 66
- shower 25, 68
- single effective subprocess 19, 84

PHOJET 26
photon
- anomalous 11, 30, 32, 61, 84, 97
- direct 52, 63, 66, 84, 89, 93
- effective parton distribution 19, 97
- fluctuations 6
- fractional energy x_γ 20, 59, 65
- fragmentation region 40
- intrinsic k_t 61, 68
- parton distribution functions 32, 82
- remnant 53, 59
- resolved 53, 59, 66
- scaled energy y 5
- structure function F_2^γ 10
- structure function F_2^γ 29
- VDM 12, 63, 84
- virtuality Q^2 5, 7, 39

photoproduction
- processes 38
- total cross section 2, 24

proton–(anti-)proton
- double parton scattering 72

– jet width 57
– particle cross section 49, 50
– underlying event 79
PYTHIA 26

QCD
– event generators
– – HERWIG 27
– – PHOJET 26
– – PYTHIA 26
– evolution equations 10
– jet and particle cross sections 16, 21
– matrix elements 14
– parton cross sections 13
– photon structure function 10
– single effective subprocess 19, 84

reference frames
– laboratory 20, 40
– parton–parton 20
– photon–proton 40

scale
– factorization 16, 22, 26
– fragmentation 17
– renormalization 15, 16, 22, 26
scattering
– diffractive 37, 49
– elastic 37
– hard 37, 39, 49, 51
– nondiffractive 24, 37
– soft 37

tagged electron events 9, 39, 40, 47
two-photon
– jet cross section 86
– jet width 58
– photon structure function 29

underlying event 54, 77, 90, 92
untagged electron events 10, 39

vector meson dominance model 12

Weizsäcker–Williams approximation 8

Springer Tracts in Modern Physics

122 **Particle Induced Electron Emission I**
With contributions by M. Rösler, W. Brauer,
and J. Devooght, J.-C. Dehaes, A. Dubus, M. Cailler, J.-P. Ganachaud
1991. 64 figs. X, 130 pages

123 **Particle Induced Electron Emission II**
With contributions by D. Hasselkamp and H. Rothard, K.-O. Groeneveld,
J. Kemmler and P. Varga, H. Winter 1992. 90 figs. IX, 220 pages

124 **Ionization Measurements in High Energy Physics**
By B. Sitar, G. I. Merson, V. A. Chechin, and Yu. A. Budagov
1993. 184 figs. X, 337 pages

125 **Inelastic Scattering of X-Rays with Very High Energy Resolution**
By E. Burkel 1991. 70 figs. XV, 112 pages

126 **Critical Phenomena at Surfaces and Interfaces**
Evanescent X-Ray and Neutron Scattering
By H. Dosch 1992. 69 figs. X, 145 pages

127 **Critical Behavior at Surfaces and Interfaces**
Roughening and Wetting Phenomena
By R. Lipowsky 1999. 80 figs. X, Approx. 180 pages

128 **Surface Scattering Experiments with Conduction Electrons**
By D. Schumacher 1993. 55 figs. IX, 95 pages

129 **Dynamics of Topological Magnetic Solitons**
By V. G. Bar'yakhtar, M. V. Chetkin, B. A. Ivanov, and S. N. Gadetskii
1994. 78 figs. VIII, 179 pages

130 **Time-Resolved Light Scattering from Excitons**
By H. Stolz 1994. 87 figs. XI, 210 pages

131 **Ultrathin Metal Films**
Magnetic and Structural Properties
By M. Wuttig 1998. 103 figs. X, Approx. 180 pages

132 **Interaction of Hydrogen Isotopes with Transition-Metals and Intermetallic Compounds**
By B. M. Andreev, E. P. Magomedbekov, G. Sicking 1996. 72 figs. VIII, 168 pages

133 **Matter at High Densities in Astrophysics**
Compact Stars and the Equation of State
In Honor of Friedrich Hund's 100th Birthday
By H. Riffert, H. Müther, H. Herold, and H. Ruder 1996. 86 figs. XIV, 278 pages

134 **Fermi Surfaces of Low-Dimensional Organic Metals and Superconductors**
By J. Wosnitza 1996. 88 figs. VIII, 172 pages

135 **From Coherent Tunneling to Relaxation**
Dissipative Quantum Dynamics of Interacting Defects
By A. Würger 1996. 51 figs. VIII, 216 pages

136 **Optical Properties of Semiconductor Quantum Dots**
By U. Woggon 1997. 126 figs. VIII, 252 pages

137 **The Mott Metal-Insulator Transition**
Models and Methods
By F. Gebhard 1997. 38 figs. XVI, 322 pages

138 **The Partonic Structure of the Photon**
Photoproduction at the Lepton-Proton Collider HERA
By M. Erdmann 1997. 59 figs. X, 118 pages

139 **Aharonov-Bohm and other Cyclic Phenomena**
By J. Hamilton 1997. 34 figs. In preparation

Printing: Mercedesdruck, Berlin
Binding: Buchbinderei Lüderitz & Bauer, Berlin